Lecture Notes in Networks and Systems 1029

The series "Lecture Notes in Networks and Systems" publishes the latest developments in Networks and Systems—quickly, informally and with high quality. Original research reported in proceedings and post-proceedings represents the core of LNNS.

Volumes published in LNNS embrace all aspects and subfields of, as well as new challenges in, Networks and Systems.

The series contains proceedings and edited volumes in systems and networks, spanning the areas of Cyber-Physical Systems, Autonomous Systems, Sensor Networks, Control Systems, Energy Systems, Automotive Systems, Biological Systems, Vehicular Networking and Connected Vehicles, Aerospace Systems, Automation, Manufacturing, Smart Grids, Nonlinear Systems, Power Systems, Robotics, Social Systems, Economic Systems and other. Of particular value to both the contributors and the readership are the short publication timeframe and the world-wide distribution and exposure which enable both a wide and rapid dissemination of research output.

The series covers the theory, applications, and perspectives on the state of the art and future developments relevant to systems and networks, decision making, control, complex processes and related areas, as embedded in the fields of interdisciplinary and applied sciences, engineering, computer science, physics, economics, social, and life sciences, as well as the paradigms and methodologies behind them.

Indexed by SCOPUS, EI Compendex, INSPEC, WTI Frankfurt eG, zbMATH, SCImago.

All books published in the series are submitted for consideration in Web of Science.

For proposals from Asia please contact Aninda Bose (aninda.bose@springer.com).

Valentina E. Balas · Kolla Bhanu Prakash ·
G. P. Saradhi Varma
Editors

Proceedings of the International Conference on Internet of Everything and Quantum Information Processing

Springer

Editors
Valentina E. Balas
Aurel Vlaicu University of Arad
Arad, Romania

G. P. Saradhi Varma
Koneru Lakshmaiah Education Foundation
Vijayawada, Andhra Pradesh, India

Kolla Bhanu Prakash
Department of Computer Science
and Engineering
Koneru Lakshmaiah Education Foundation
Guntur, Andhra Pradesh, India

ISSN 2367-3370 ISSN 2367-3389 (electronic)
Lecture Notes in Networks and Systems
ISBN 978-3-031-61928-1 ISBN 978-3-031-61929-8 (eBook)
https://doi.org/10.1007/978-3-031-61929-8

This Springer imprint is published by the registered company Springer Nature Switzerland AG
The registered company address is: Gewerbestrasse 11, 6330 Cham, Switzerland

Preface

This volume constitutes the Proceedings of the International Conference on Internet of Everything and Quantum Information Processing (IEQIP-2023), held on 24–25 November 2023 in Andhra Pradesh, India. This edition was organized in hybrid mode by the Department of Computer Science and Engineering, Koneru Lakshmaiah Education Foundation, and sponsored by the Department of Science and Technology, India.

The Internet of Things (IoT) and Quantum Computing are technological phenomena that are shaping businesses and industries. Moreover, integrating IoT and Quantum Computing solutions evolves ecosystems to offer optimum commercial practices. The emergence of these two phenomena offers revolutionary advancements. Further, Quantum Computing introduces avant-garde encryption algorithms for resilient security solutions in IoT systems. Papers presented at this conference showed how the Internet of Everything (IoE) combines sensing, computation, information extraction, and communication functionalities together in a device. IoE allows different electronic devices with different capabilities to sense the environment and to communicate for data exchange. Further, the challenge now is to identify, monitor, and ensure quality in IoT security. Moreover, there is a significant rise in data and information transfer due to a rise in the number of devices. Hence, there is a requirement for responsive, scalable, analytical, and efficient solutions.

This conference provided a premier venue for showcasing top-notch, original research, game-changing inventions, and captivating insights into artificial intelligence, quantum engineering, machine learning, deep learning, data science, and the Internet of Everything.

The conference papers included in this volume contain topics in the following areas of research:

- Artificial Intelligence
- Quantum Computing
- Internet of Things, Cloud Computing, and Data Science
- Machine learning
- Deep learning
- Image, text, and signal processing
- Disruptive technologies

In IEQIP 2023, we had many eminent Keynote Speakers: Dr. Kathleen Kramer (USA), Dr. Rajkumar Buyya (Australia), Dr. Edwin Chong (USA), Dr. Tran Duc-Tan (Vietnam), Dr. Deepak Garg (India), Rut Lineswala (India), Jash Kumar (India).

We are thankful to all the authors from different countries who have submitted papers to keep the quality of the IEQIP 2023 conference at high levels. The editors of this book would like to acknowledge all the authors, session chairs, and reviewers for their contributions. We have received invaluable help from the members of the International Program Committee and the Chairs responsible for different aspects of the event.

For their help with the IEQIP 2023 edition, we thank all the University Administration, Faculty, Scholars, and Students for customizing the registration of conference participants and all other required arrangements.

Our special thanks go to Dr. Janus Kacprzyk (Editor-in-Chief, Springer, Lecture Notes in Networks and Systems) for the opportunity to organize this guest-edited volume.

We are grateful to Springer, especially to Dr. Thomas Ditzinger (Senior Editor, Applied Sciences & Engineering Springer-Verlag), for the excellent collaboration, patience, and help during the evolution of this volume.

We hope this volume will provide helpful information to professors, researchers, and graduate students in the area of quantum computing and IoT, and all will find this collection of papers inspiring, informative, and valuable. We also hope to see you at a future IEQIP event.

Valentina E. Balas
Kolla Bhanu Prakash
G. P. Saradhi Varma

Technical Program Committee Members

Andrzej Ruciński	Internet of Things Laboratory-IOT Lab, University of New Hampshire
Durgesh Kumar Mishra	Department of CSE, Sri Aribundo Institute of Technology, Indore, Madhya Pradesh
Srinivas Mukkamala	Department of Computer Science, USA
Deepak B. Phatak	Department of CSE, IIT-Bombay
Narayanan Kulathuramaiyer	Institute of Social Informatics & Technological Innovations University, Malaysia Sarawak, Malaysia
Dharam Singh Jat	Faculty of Computing and Informatics, Namibia University of Science and Technology, SA
Arvind Shah	Georgia Southwestern State University, Americus
Satyen Parikh	Cambodia Operations, Cambodia
Sharul Kamal Abdul Rahim	Universiti Teknologi Malaysia
Jane Labadin	Universitiy Malaysia Sarawak Malaysia
M. Hassaballah	Department of CS, South Valley University, Luxor, Egypt
K. V Ramana	Department of CSE, JNTUK
H. R. Vishwakarma	VIT University
D. Janakiram	IITM, Chennai
Chandrashekhar Sahasrabudhe	ACM India
S. C. Satapathy	Professor in CSE, Director Research, KIT
Vivek S Deshpande	Technical University of Sofia, Bulgaria, Europe
Anicia Peters	Namibia University of Science and Technology, SA
S. K. Hafizul Islam	IIIT, Kalyani
Tanmay De	National Institute of Technology, Durgapur, West Bengal
Saptarshi Ghosh	Dept. of CSE, Indian Institute of Technology, Kharagpur
Somnath Bhattyacharya	Computer Science and Engineering, Assam University, Silchar, Assam
Indrajit Banerjee	Department of Information Technology, IIEST Shibpur, W. B.
Santanu Phadikar	Maulana Abul Kalam Azad University of Technology, West Bengal
B. Acharya	Department of ETC, NIT, Raipur, India
Satchidananda Dehuri	Fakir Mohan University, Balasore, Odisha, India
G. T. Chandra Sekhar	Sri Sivani College of Engineering, Srikakulam, Andhra Pradesh
P. Vijaykumar	UCE, Tindivanam, Chennai

To my husband, Marius, For your unwavering support and belief in me.

Valentina

Acknowledgment

I would like to say thank you to the Almighty and my parents for the endless support, guidance, and love through all stages. I dedicate this book to my parents, family members, and my wife.

I would like to specially thank Sri. Koneru Satyanarayana, President, K. L. University, India, for his continuous support and encouragement throughout the preparation of this book.

I would like to express my gratitude towards all who supported, shared, talked things over, read, wrote, offered comments, allowed us to quote remarks and assisted in editing, proofreading and design, through this book journey. We pay our sincere thanks to the open data set providers.

I am grateful to Springer Publishing team, who showed us the ropes to creating this book. Without that knowledge I would not have ventured into starting this conference and preparing the proceedings book, which ultimately led to this. Their trusting in us, their guidance and providing the necessary time and resources, gave us the freedom to manage this book.

Last, but definitely not least, I'd like to thank my readers, who gave us their trust and hope this work inspires and guides them.

Valentina E. Balas
Kolla Bhanu Prakash
G. P. Saradhi Varma

Contents

About the Editors

Valentina E. Balas is currently Full Professor in the Department of Automatics and Applied Software at the Faculty of Engineering, "Aurel Vlaicu" University of Arad, Romania.

She holds a Ph.D. Cum Laude, in Applied Electronics and Telecommunications from Polytechnic University of Timisoara. Dr. Balas is the author of more than 400 research papers in refereed journals and International Conferences. Her research interests are in Intelligent Systems, Fuzzy Control, Soft Computing, Smart Sensors, Information Fusion, Modeling and Simulation.

She is the Editor-in Chief to International Journal of Advanced Intelligence Paradigms (IJAIP) and to International Journal of Computational Systems Engineering (IJCSysE), member in Editorial Board member of several national and international journals and is evaluator expert for national, international projects and PhD Thesis.

Dr. Balas is the Head of Intelligent Systems Research Centre in Aurel Vlaicu University of Arad.

Dr. Balas is member in the specialized committee "Computers, Information Technology, and Systems Engineering" of the National Council for Attestation of University Titles, Diplomas, and Certificates (CNATDCU) for the 2024-2028 term.

She served as General Chair of the International Workshop Soft Computing and Applications (SOFA) in ten editions organized in the interval 2005-2022 and held in Romania and Hungary.

Dr. Balas participated in many international conferences as Organizer, Honorary Chair, Session Chair, member in Steering, Advisory or International Program Committees and Keynote Speaker. Recently she was working on a national project with EU funding support: BioCell-NanoART = Novel Bio-inspired Cellular Nano-Architectures - For Digital Integrated Circuits, 3M Euro from National Authority for Scientific Research and Innovation.

She is a member of European Society for Fuzzy Logic and Technology (EUSFLAT), member of Society for Industrial and Applied Mathematics (SIAM) and a Senior Member IEEE, member in Technical Committee – Fuzzy Systems (IEEE Computational Intelligence Society), chair of the Task Force 14 in Technical Committee – Emergent Technologies (IEEE CIS), member in Technical Committee – Soft Computing (IEEE SMCS). She is member in the Committee of IEEE Romania Section as Volunteers Training Coordinator and vice chair of IEEE Computational Intelligence Society Chapter – CIS 11. During the interval 2021-2022 she was a member of IEEE European Public Policy Committee Working Group on ICT. From May 2023 Dr. Balas is associate member of Romanian Academy of Scientists.

Dr. Balas was past Vice President (awards) of IFSA - International Fuzzy Systems Association Council (2013-2015), is a Joint Secretary of the Governing Council of Forum for Interdisciplinary Mathematics (FIM), - A Multidisciplinary Academic Body, India.

She is the recipient of the "Tudor Tanasescu" Prize from the Romanian Academy for contributions in the field of soft computing methods (2019) and "Stefan Odobleja" Prize from Romanian Academy of Scientists (2023 and 2024) and Diploma – Section Information Technology from The General Association of the Engineers in Romania (AGIR) 2023.

Dr. Kolla Bhanu Prakash is working as Professor, Associate Dean [R&D], and Research Group Head for A.I & Data Science Research Group at KLEF. He is Adjunct Professor at Taylors University, Malaysia. He has published 150+ research papers in international and national journals and conferences. He published Scopus and SCI publications 89 with an H-Index of 16 and a total citations of 1085. He is Reviewer for several international journals, including IEEE, Elsevier, Springer, Wiley, Taylor, and Francis, and served as Reviewer, Keynote Speaker, and TPC Member for several International conferences. His research interests include deep learning, data science, and quantum computing. He received the Best Researcher Award 4 times. He authored 2 books and edited 12 books with internationally reputed publishers like IEEE, Elsevier, Springer, River, CRC, Degryuter, and Wiley. He also published 15 patents, 2 patent grants, and 1 Copyright. He is guiding 4 Ph.D. scholars. Under his guidance, 3 Ph.D. students were awarded doctoral degrees. He is IEEE Senior Member, Fellow-ISRD, and Treasurer—ACM Amaravathi Chapter, India, LMISTE, MIAENG, SMIRED. He is Series Editor of "Next Generation Computing & Communication Engineering" for Wiley Publishers. He is Series Editor of "Industry 5.0: Artificial Intelligence, Cyber-Physical Systems, Mechatronics, and Smart Grids" for CRC publishers. He is NAAC Accessor. He is listed in the world's top 2% of researchers cited by Stanford University and Elsevier for 2021. He received a grant of 27 lakhs from Meity for establishing a Quantum Computing Research lab [QCAL] at KLEF and a grant from DST SERB worth 3 lakhs for organizing a workshop and conference in Quantum Computing. He delivered many Keynote speeches and expert talks worldwide in quantum computing, machine learning, and deep learning.

Prof. G. P. S. Varma Vice-Chancellor, KLEF, is one of the most widely experienced leaders in Indian higher education, known for his commitment to expanding student opportunity, catalyzing academic innovation, and encouraging the university's civic engagement and service to society. He adorned the position of Chairman, ISTE (Indian Society for Technical Education)—AP State, TSEMCET Test Committee Member-2021 nominated By Telangana State Govt, APEAMCET Admission Committee Member in 2016 by Andhra Pradesh State Council of Higher Education, Govt. of Andhra Pradesh. He has been a very farsighted Peer Team Visit Member for the National Assessment and Accreditation Council (NAAC) and an expert Committee Member for University Grants Commission (UGC) Autonomous Visits.

In the capacity of Board Member of IUCEE (Indo US Collaboration for Engineering Education), Board Member of GEDC (Global Engineering Deans Council), Membership Chair of ASEE (American Society for Engineering Education) Indian Chapter, Executive Body Member for Andhra Pradesh Private Engineering Colleges Managements Association (APPECMA) and Former Board Member for Storage Area Networks Industrial Association, the visionary professor advanced the unrelenting commitment to excellence in teaching, research, and public service and fostered significant collaborations across the globe.

He has received the Top 100 Educators 2015 award by the International Biographical Centre, Cambridge, England.

He has been honored with the 2000 Outstanding Intellectuals of the 21st Century medal by the International Biographical Centre, Cambridge, England. He has received the Vice Chancellor Excellence Award-2022 by AICTE—EduSkills. He has received the Lifetime Achievement Award 2021–2022 from the Bharat Education Excellence Awards, the All India Level most significant award ceremony for education and research.

During his pristine career, he held visiting delegations at universities in Israel, USA, Germany, Switzerland, Norway, Poland, Austria, Dubai, and China to submit a report on how to enhance the quality of education as per the Indian scenario.

Under his guidance, 22 doctoral students were awarded. He published 6 books, 9 patents, 38 National, and 153 International research papers in various reputed International and National Journals. While guiding KLEF through successful leaps, he brought to fruition the most ambitious projects funded by AICTE, UGC, DST, and NITI Aayog in the university's history.

He, the audacious leader, guided the university's efforts in undergraduate and graduate education, research initiatives, international and industrial partnerships, and strategic planning while playing an integral role in soaring the prestige of KLEF exponentially in the global arena of world-class education leaders.

Enhancing Cloud Security and Resource Management: A Comprehensive Review

B. Subramanya Anil Kumar(✉) and Basant Sah

Department of CSE, Koneru Lakshmaiah Education Foundation, Guntur, AP, India
batchuanilkumar@gmail.com, basantbitmtech2008@kluniversity.in

Abstract. The rapid proliferation of cloud computing has ushered in a new era of convenience and scalability for organizations and individuals alike. However, this convenience also comes with a host of security and resource management challenges that demand innovative solutions. This review paper embarks on a comprehensive exploration of three pivotal objectives aimed at fortifying the cloud ecosystem. This paper also offers a thorough perspective on the changing environment of cloud security and resource management by synthesising and analysing the most recent developments in these fields. Researchers, practitioners, and other interested parties may all benefit from it since it provides insights that can spur further innovation and growth in the rapidly developing area of cloud computing.

Keywords: Cloud Security · Data Integrity · User Authentication · Resource Optimization · Encryption · Virtual Machine (VM) Management

1 Introduction

Cloud computing has enabled businesses to streamline operations, reduce infrastructure costs, and adapt to ever-changing market demands [1]. However, this transformational shift towards the cloud also brings forth a new set of challenges, notably in the realms of security and resource management.

Strong security measures become crucial in this circumstance to ensure the availability, confidentiality, and integrity of data. Organizations must not only safeguard their data but also demonstrate compliance with regulations like GDPR, HIPAA, or industry-specific standards, which often come with strict security requirements [2]. These commitments requires the ability to allocate resources dynamically based on application needs, user demands, and performance objectives. These mechanisms enable cloud providers to match available resources with application workloads effectively, optimizing resource utilization while maintaining acceptable levels of performance and security [3].

This review paper embarks on a comprehensive exploration of these critical facets of cloud computing – security and resource management. It develops into the development and evaluation of cutting-edge models for multi-level user authentication, data integrity verification, and VM failure detection. By synthesizing and analyzing the latest advancements in these domains, this paper aims to provide a holistic perspective on the evolving landscape of cloud security and resource management.

V. E. Balas et al. (Eds.): IEQIP 2023, LNNS 1029, pp. 1–6, 2024.
https://doi.org/10.1007/978-3-031-61929-8_1

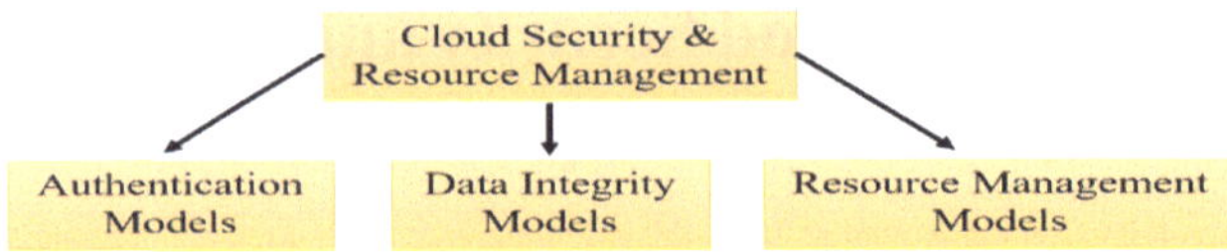

Fig. 1. Cloud Security & Resource Management

The Fig. 1 acts as a visual guide to our comprehensive review, providing a clear and structured overview of the intricate relationship between cloud security and resource management. This figure serves as the central reference point that unifies our exploration of these critical facets of cloud computing, emphasizing their deeply interwoven nature. At its core, the figure places cloud security as the central pillar, representing the bedrock upon which our entire study is founded. Cloud security is the linchpin that ensures the confidentiality, integrity, and availability of data and services in the cloud environment. Radiating outwards from the central core, we encounter three distinct segments, each encapsulating a vital domain of investigation: Authentication Models (Fig. 2), Data Integrity Models (Fig. 3), and Resource Management Models (Fig. 4).

Cloud User Authentication: User authentication is a key component of security in the world of cloud computing. It is crucial to strengthen this component against changing threats since it acts as the entry point to cloud resources [4].

Traditional Authentication Challenges: Despite their widespread usage, traditional authentication systems are rife with flaws [5]. In the ever-evolving landscape of cybersecurity, the trustworthiness of password-based security measures has come under scrutiny due to a plethora of critical vulnerabilities. In light of the previously discussed work in the domain of cloud computing and security, it is imperative to delve deeper into the pressing challenges associated with password-based authentication.

Comparative study with current ways: The Aytaj Badirova et al. paper explores multi-level user authentication, data integrity verification, and VM failure detection models, all with the goal of boosting cloud security and resource management, build on this by laying the groundwork [6].

In their study, P. Hari Kumar et al. focus on the significance of cloud data storage within the context of cloud providers like Amazon Web Services, Google Cloud Platform, and Microsoft Azure. The proposed system leverages multi-copy storage across multiple cloud servers, enhancing data security and reducing dependency on a single provider.

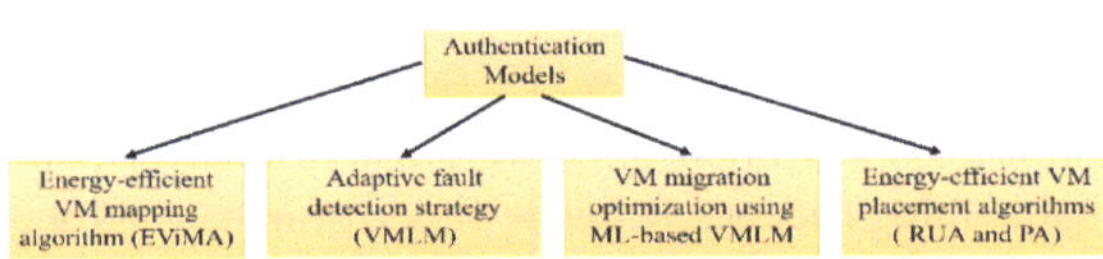

Fig. 2. Authentication Models

Authentication Models is crucial for understanding this work core concepts (see Fig. 2). It provides a structured overview of authentication models, including details about each model's purpose and benefits. Energy-efficient VM mapping algorithm

(EViMA), is highlighted, showcasing its proposal and achievement of better scheduling performance, improved energy consumption, and a reduced makespan. Adaptive fault detection strategy (VMLM), is presented with clarity, offering insights into its purpose and benefits within the realm of cloud security. VM migration optimization using machine learning-based VMLM, is depicted, shedding light on its role in reducing VM migrations, lowering energy consumption, and achieving a significant reduction in fault ratio. Energy-efficient VM placement algorithms (RUA and PA), are showcased, emphasizing their introduction and demonstration of trade-offs, particularly RUA's role in reducing SLA violations and preventing overloading.

2 Data Integrity Verification in the Cloud

This section explores the several facets of data integrity verification and how it contributes to cloud security.

Issues with Maintaining Data Integrity: The cloud environment presents a number of issues with keeping data integrity [7]. Threats to data include insider threats, unauthorised access, data corruption during transit or storage, and others. Data may be dispersed across several servers and locations, making its tracking and validation difficult [8].

Strong Encryption and Decryption methods: Data integrity verification spans the whole lifespan of data in the cloud thanks to lifecycle management [9]. Monitoring and verifying data integrity at each of these stages are critical to preventing tampering, unauthorized changes, or data loss [10].

Fig. 3. Data Integrity Models

Data Integrity Models is pivotal for comprehending the core concepts of this work (see Fig. 3). The figure offers insights into various data integrity models, including details about each model's purpose and benefits. For instance, 'IoT Access Control and Data Integrity with MLS-ABAC scheme and Lightweight Functions' is highlighted,. 'Enhancing Cloud Storage Security and Reducing Vendor Lock-in with Dynamic Identity-Based System' is presented, shedding light on its role in providing a multi-copy storage system for flexibility and security.

Case Studies and Use Cases: To illustrate the practical application of data integrity verification in the cloud, this section provides real-world case studies and use cases. Data integrity issues are addressed by Mani Deep Karumanchi et al. in cloud-based Internet of Things (IoT) applications, especially in the dynamic IoT data environment. Experimental findings show that their improved supply chain data integrity model outperforms conventional methods in terms of integrity bit change, runtime, and encoding average runtime [11].

Importance of VM Failure Prevention: VMs serve as the foundation for hosting applications and services and are the basic building blocks of cloud infrastructure [12].

Task Rescheduling for Resource Optimisation: Task rescheduling turns into a crucial tactic to avoid underutilizing cloud resources in the case of VM failure or the prediction of failure [13, 14].

Real-world Implementation and Results: These implementations can showcase the adaptability of these approaches in different cloud environments, highlighting their effectiveness in mitigating VM failures and optimizing resource allocation.

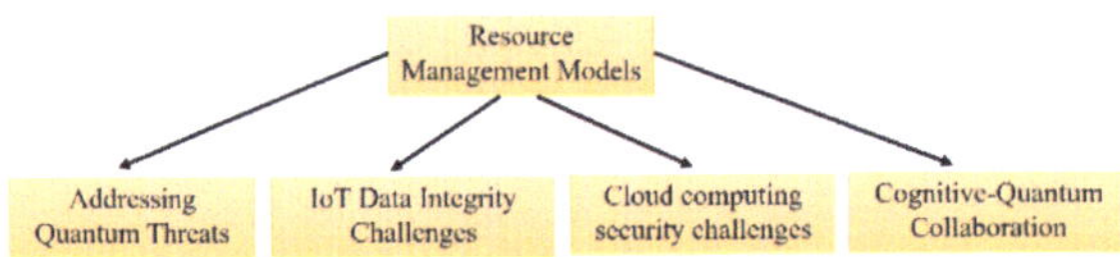

Fig. 4. Resource Management Models

Resource Management Models are instrumental in comprehending the core concepts of this work (see Fig. 4). The figure offers insights into various resource management models, including details about each model's purpose and benefits. It introduces the proposed CDIVM model for cloud data integrity. 'IoT Data Integrity Challenges: Non-Linear Integrity Model with Dynamic Key Size and Cipher Text Policy' is presented, shedding light on its novel non-linear integrity computational model for cloud-based IoT, dynamic key size, and improved data integrity efficiency. 'Cognitive-Quantum Collaboration' is showcased as well, highlighting its cognitive-quantum collaboration for NCCC security, boasting high throughput, and successful CMMLA implementation.

Essam H. Houssein et al. delve into cloud task scheduling, a critical component of cloud computing, focusing on meta-heuristic algorithms. They provide an overview of traditional and heuristic scheduling methods and then systematically categorize various meta-heuristic techniques based on scheduling problem attributes. Ali Belgacem et al. introduce VMLM, a machine learning-driven solution to optimize virtual machine (VM) migrations in cloud computing. It leverages machine learning to make real-time migration decisions, adapting to dynamic resource allocation. Algorithms, making it a pioneering approach to VM migration in the cloud. Future research may explore further enhancements to the learning model [15].

3 Integration and Synergy

The Interplay between User Authentication, Data Integrity, and VM Management: Data integrity verification safeguards data against tampering and unauthorized access, preserving its trustworthiness. VM management orchestrates the underlying infrastructure, optimizing resource allocation for seamless service delivery.

Achieving Comprehensive Cloud Security and Efficiency: The integration of these domains culminates in the realization of comprehensive cloud security and efficiency. When multi-level user authentication is paired with robust data integrity verification, the security perimeter is fortified [16].

Use Cases Demonstrating the Synergy: These cases will showcase real-world scenarios where the integration of these domains has yielded tangible benefits.

4 Comparative Analysis

This experimental finding focus on measurable improvements confirms the applicability and effect of the suggested solutions in the real world, highlighting their importance for both cloud providers and consumers.

Resource management models and VM failure detection models aim to optimize resource utilization while minimizing energy consumption and the multifaceted nature of Cloud computing.

The comparative analysis underscores the collective contribution of the reviewed literature in advancing the field of cloud technology, offering practical insights and solutions that fortify data security, resource management, and overall efficiency.

5 Challenges and Future Directions

Continuous research, cloud security, and resource management are key considerations. Cyberthreats, data privacy in multi-cloud systems, and dynamic infrastructure resource distribution are challenges. Real-time threat detection, multi-cloud data security encryption, and adaptive resource allocation should be researched. Standard APIs also improve cloud system compatibility.

6 Conclusion

This extensive overview addresses important facets of cloud security and resource management in the changing world of cloud computing. We concentrate on three main goals: data integrity verification, VM failure detection, and multi-level user authentication. Our findings have broad ramifications for how cloud computing will develop in the future. To improve security and access control, we draw attention to the drawbacks of conventional user authentication and present multi-level unique key-based validation. Data integrity verification places a high emphasis on lifecycle management and secure encryption methods. For continuous operation and the most effective use of resources, VM failure detection and rescheduling become essential. These discoveries alter cloud resource management and security, providing better defence, compliance, and effectiveness. Increased security is achieved by using multi-level user authentication to defend against breaches. Verifying data integrity makes sure it hasn't been tampered with, which is essential for trust. Resource allocation is optimized via proactive VM failure detection and rescheduling. Concludingly, these approaches improve cloud resilience, heralding a revolutionary digital future.

References

1. Huang, H.Y., Ku, E.C., Chen, C.D.: Cloud infrastructure enhancing product competitive advantage of tourism SMEs on online consumption values of tourists. Bus. Process. Manag. J. **28**(4), 1146–1163 (2022)
2. Paricherla, M., Babu, S., Phasinam, K., Pallathadka, H., Zamani, A.S., Narayan, V., Shukla, S.K., Mohammed, H.S.: Towards development of machine learning framework for enhancing security in Internet of Things. Security and Communication Networks **2022**, 4477507 (2022)
3. Buyya, R., Garg, S.K., Calheiros, R.N.: SLA-oriented resource provisioning for cloud computing: challenges, architecture, and solutions. In: 2011 International Conference on Cloud and Service Computing, pp. 1–10. IEEE 2011
4. Lombardi, F., Di Pietro, R.: Secure virtualization for cloud computing. J. Netw. Comput. Appl. **34**(4), 1113–1122 (2011)
5. Zheng, X., Julien, C., Kim, M., Khurshid, S.: On the state of the art in verification and validation in cyber physical systems. In: The University of Texas at Austin, The Center for Advanced Research in Software Engineering, Tech. Rep. TR-ARiSE-2014-001, pp. 1485 (2014)
6. Badirova, A., Dabbaghi, S., Moghaddam, F.F., Wieder, P., Yahyapour, R.: An optimized single sign-on schema for reliable multi-level security management in clouds. In: 2021 8th International Conference on Future Internet of Things and Cloud (FiCloud), pp. 42–49. IEEE (2021)
7. Zhou, L., Fu, A., Yu, S., Su, M., Kuang, B.: Data integrity verification of the outsourced big data in the cloud environment: a survey. J. Netw. Comput. Appl. **122**, 1–15 (2018)
8. Chen, Y., Venkatesan, R., Cary, M., Pang, R., Sinha, S. and Jakubowski, M.H.: Oblivious hashing: a stealthy software integrity verification primitive. In: Information Hiding: 5th International Workshop, IH 2002 Noordwijkerhout, The Netherlands, October 7-9, 2002 Revised Papers 5, pp. 400–414. Springer Berlin Heidelberg (2003). https://doi.org/10.1007/3-540-36415-3_26
9. Ahamad, S., Roshan, A., Lourens, M., Shekher, V., Joshi, K. and Alanya-Beltran, J.: The critical role played by big data management in effectively addressing the security and overall privacy concerns through correlation analysis. In: 2022 2nd International Conference on Advance Computing and Innovative Technologies in Engineering (ICACITE), pp. 130–134. IEEE (2022)
10. Wei, P., Wang, D., Zhao, Y., Tyagi, S.K.S., Kumar, N.: Blockchain data-based cloud data integrity protection mechanism. Futur. Gener. Comput. Syst. **102**, 902–911 (2020)
11. Karumanchi, M.D., Sheeba, J.I., Devaneyan, S.P.: Integrated Internet of Things with cloud developed for data integrity problems on supply chain management. Meas. Sens. **24**, 100445 (2022)
12. Dawoud, W., Takouna, I., Meinel, C.: Infrastructure as a service security: challenges and solutions. In: 2010 the 7th International Conference on Informatics and Systems (INFOS), pp. 1–8. IEEE (2010)
13. Sana, M.U., Li, Z.: Efficiency aware scheduling techniques in cloud computing: a descriptive literature review. PeerJ Computer Science **7**, e509 (2021)
14. Houssein, E.H., Gad, A.G., Wazery, Y.M., Suganthan, P.N.: Task scheduling in cloud computing based on meta-heuristics: review, taxonomy, open challenges, and future trends. Swarm Evol. Comput. **62**, 100841 (2021)
15. Belgacem, A., Mahmoudi, S., Ferrag, M.A.: A machine learning model for improving virtual machine migration in cloud computing. J. Supercomput. **79**, 1–23 (2023)
16. Musonda, C., Monica, M.K., Nyirenda, M., Phiri, J.: Security, privacy and integrity in internet of things–a review. In: Proceedings of the ICTSZ International Conference in ICTs, pp. 148–152 (2019)

A Novel Approach Integrating Autoencoders and ESMOTE-GAN for Credit Card Fraud Detection

Sai Kiran Pasupuleti[1,2], Tirapathi Reddy Burramukku[3], Sasank Sai Rayapati[1,2](✉), Hari Krishna Vamsi Yarlagadda[1,2], Krishna Priya Suda[1,2], and Sai Snigdha Valluripalli[1,2]

[1] Prasad V. Potluri Siddhartha Institute of Technology, Chalasani Nagar, Kanuru, Vijayawada, India
psaikiran@pvpsiddhartha.ac.in, sasankrayapati777@gmail.com, yharikrishnavamsi28@gmail.com

[2] Springer Heidelberg, Tiergartenstr. 17, 69121 Heidelberg, Germany

[3] Koneru Lakshmaiah Education Foundation, Green Fileds, Vaddeswaram, AP, India

Abstract. An innovative Credit Card Fraud Detection Algorithm that merges two potent techniques, Autoencoders and Ensemble Synthesized Minority Oversampling Techniques using GANs (ESMOTE- GAN), to effectively address the pressing issue of credit card online payment fraud. The fraud detection, traditional models encounter significant challenges due to the skewed nature of transaction datasets, with the majority of entries representing legitimate transactions. Our approach tackles this challenge head-on by elevating the quality of feature representation and diminishing noise through the application of Autoencoders. By reducing the dimensionality and highlighting meaningful patterns within the data, Autoencoders enable our algorithm to capture essential information while ignoring extraneous variables. Furthermore, we strategically employ ESMOTE-GAN to rectify dataset imbalances. This entails generating synthetic minority-class samples, thus rectifying the skewed distribution without compromising on model accuracy and improvements of performance and the detection rate. Furthermore, our technique was rigorously evaluated by comparing it against a range of classifiers, such as Random Forest (RF), k-Nearest Neighbours (KNN), Support Vector Machines (SVM) and XG Boost. We streamlined the decision-making process by integrating the probability outputs from our trained classifiers through a weighted voting system. The results clearly indicate substantial enhancements in both overall performance and detection rates. Therefore, our model's heightened detection capabilities not only enhance security but also lead to a significant reduction in the resources needed for manual analysis, presenting a more efficient and cost-effective solution.

Keywords: Credit Card Fraud · Autoencoder · ESMOTE · over-fitting · Random Forest

V. E. Balas et al. (Eds.): IEQIP 2023, LNNS 1029, pp. 7–16, 2024.
https://doi.org/10.1007/978-3-031-61929-8_2

1 Introduction

The importance of credit card fraud detection for ensuring financial security has grown significantly in recent years. With the progression of technology, fraudsters have also evolved their strategies, necessitating financial institutions to utilize sophisticated techniques to safeguard their clients. Credit card fraud is the act of misusing or using credit card information without the cardholder's consent or knowledge, leading to financial losses for both individuals and institutions. In order to address this issue, a range of fraud detection methods are utilized, frequently leveraging machine learning and data analysis. These techniques entail examining transaction data to identify abnormal patterns, anomalies, and indications of suspect conduct. Examples of anomalies could encompass a typical purchasing locations, transaction frequencies, or irregular spending habits. Machine learning algorithms can be trained to identify these patterns and signal alerts when they are discovered. As Cybercriminals continually develop new tactics to exploit vulnerabilities, traditional methods of identifying fraudulent activities fall short. To counter these evolving threats, this research introduces a novel and cutting-edge approach that combines Autoencoders and ESMOTE-GAN (Enhanced Synthetic Minority Over-sampling Technique - Generative Adversarial Network).This innovative fusion not only enhances the accuracy of fraud detection but also addresses the challenge of imbalanced datasets by generating synthetic samples of the minority class, thereby enabling a more effective and efficient defence against fraudulent credit card transactions [1].

The reinforcement of fraud prevention techniques is amplified by the incorporation of NFC technology for user re- authentication and theutilization of the Internet of Things for dual transaction verification. The study delves into the efficiency of feature selection and dataset division for training and testing across methodologies like Random Forest, Bayesian Network, and MultilayerPerceptron [2]. As technological advancements continue to drive the increase in credit card users, the incidence of fraudulent activities has grown in parallel, presenting a formidable challenge [3]. The increasing number of people using credit cards is accompanied by an increase in the complexity and sophistication of fraudulent activity [4].

The advent of the internet and the widespread adoption of online shopping have revolutionized financial transactions, particularly the use of credit cards for online purchases. However, this surge in online financial activity has also seen a concurrent increase in incidents of financial fraud. In 2018 alone, a staggering $24.26 billion was lost globally to credit card fraud [5]. By analysing historical transaction data and creating models based on known fraud patterns, these organizations can assess the likelihood of new transactions being fraudulent [6] (Fig. 1).

One important and well-validated technique for creating synthetic samples along a line segment is the Synthetic Minority Oversampling Technique (SMOTE), which has undergone several successful iterations. In contrast, Generative Adversarial Networks (GANs) were initially developed for image generation through adversarial training. The key distinction lies in their underlying principles: SMOTE primarily relies on local information for generating samples, while GAN methods learn from the global class distribution, making them distinct in their approach to data augmentation and generation [7].

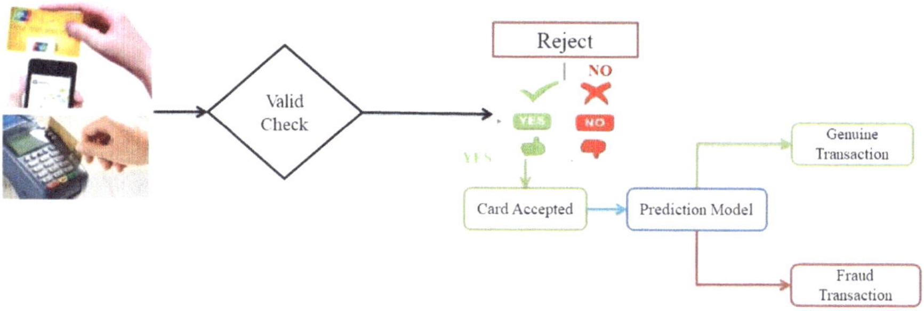

Fig. 1. Fundamental Steps in Detecting payment card fraud

2 Literature Survey

The development of an advanced fraud detection solution employing cutting-edge Machine Learning techniques. At the core of our algorithms, which encompass XG Boost, Support Vector Machine, Decision Tree, Random Forest, and Logistic Regression, lays the foundation for a resilient fraud detection framework. To further enhance efficiency, a Convolutional Neural Network (CNN) is introduced, leveraging its layered architecture to ensure precise detection. The best possible CNN model configuration is determined by a thorough empirical study that takes into account hidden layer counts, epochs, and practical applications. Additionally, important performance parameters such as F1 score, accuracy, precision, and recall are adjusted. The fact that the Area Under Curves (AUC) reach values of 99.9%, 85.71%, 93%, and 98% highlights how effective our approach, and ROC curves constructed from confusion matrices offer a visual representation of the results [8]. The Credit Card Fraud Detection System is a pivotal tool designed to identify and combat fraudulent activities within credit card transactions. This project leverages a diverse range of machine learning models and algorithms, with a particular focus on logistic regression. Through a meticulous selection process, the algorithm with the highest accuracy is employed to detect fraudulent transactions, with the logistic regression algorithm exhibiting an impressive accuracy rate [9].

Credit card fraud, a prominent subset of financial fraud, has become particularly concerning as online transactions proliferate [10, 11]. As technology advances, credit card usage grows, leading to increased opportunities for fraud. Global credit card fraud losses, which were around 28.65 billion in 2019, are projected to reach approximately 32.96 billion by 2023. To tackle this, we employed deep learning techniques, specifically the OSCNN model (Over Sampling with Convolutional Neural Network), alongside MLP, to detect and prevent credit card fraud. Comparative results demonstrate the OSCNN model's superior performance, achieving 98% accuracy, emphasizing its effectiveness in safeguarding legitimate transactions [12]. Logistic regression as a machine learning tool for detection, particularly in imbalanced datasets wherefraudulent cases are limited. We mitigate imbalance through under-sampling and over- sampling techniques, yielding a robust 94% accuracy in identifying fraudulent transactions. Moreover, our feature analysis underscores the significance of variables like transaction amount, origin country, and timing in fraud prediction. This study underscores logistic regression's potential in fraud detection, offering insights for the enhancement of financial institution security and trustworthiness [7]. This analysis delves into the practical operational aspects of FDS, presenting a formal depiction of the classification problem at hand. It underscores the vital alert-feedback loop, the conduit for current supervised samples to refine the classifier. Unlike conventional performance metrics, this study asserts that, in a real-world FDS, the precision of reported alerts holds utmost significance as investigators can only scrutinize a limited number of alerts [13].

The main contributions of this work are:

Innovative Algorithm: The introduction of a novel Credit Card Fraud Detection Algorithm that combines Autoencoder for feature extraction and dimensionality reduction, significantly improving feature representation.

ESMOTE-GAN for Imbalanced Data: The use of (ESMOTE-GAN) to tackle class imbalance in transaction datasets, ensuring dataset balance without over-fitting.

XG Boost Performance: The development of a superior XG Boost algorithm, which outperforms existing methods in terms of accuracy, providing a reliable tool for robust fraud detection.

3 Proposed Augmentation Methodology

3.1 Data Set

This dataset is a synthetic representation of credit card transactions spanning from January 1, 2019, to December 31, 2020, encompassing a mix of both genuine and fraudulent transactions. It simulates the interactions of 1000 credit card customers with a diverse set of 800 merchants. While this dataset is not derived from real-world data, it serves as a valuable tool for testing and developing credit card fraud detection models and assessing transaction behaviour patterns in a controlled environment.

3.2 Auto Encoder with ESMOTE-GAN

Combining auto-encoders with ESMOTE-GAN for credit card fraud detection can be a powerful approach to address the imbalanced nature of the dataset while also learning meaningful representations of the data [1].

Figure 2 depicts an auto-encoder-based optimal neural network architecture is a novel model design that uses auto- encoders to enhance and simplify the structure of the neural network. Autoencoders as neural networks themselves, are employed for feature extraction and dimensionality reduction. By integrating this technique, the architecture achieves more efficient and effective information processing (Fig. 3).

An Autoencoder enabled ESMOTE-GAN is a novel approach that combines the power of auto encoders and Ensemble Synthesized Minority Oversampling techniques with Generative Adversarial Networks (GANs) to address imbalanced datasets effectively. Autoencoders, aform of neural network, are employed to extract essential features from the data, enabling meaningful data representations. These representations are then

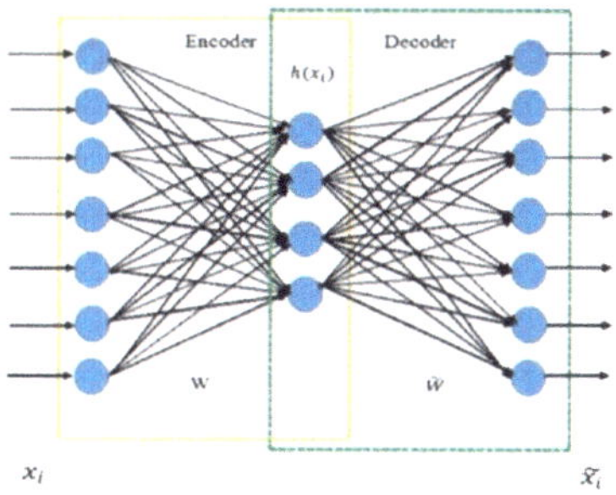

Fig. 2. Auto Encoder based Optimal Neural Network Architecture

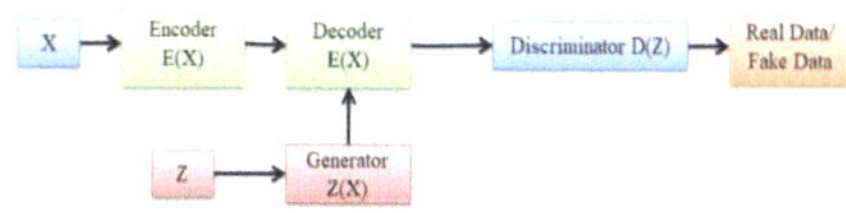

Fig. 3. Auto Encoder enabled ESMOTE GAN

leveraged in the ESMOTE-GAN framework, which generates synthetic data, specifically focusing on the minority class [14]. By merging the enhanced features from the auto encoder with synthetic data, this approach creates a more balanced and informative dataset for training machine learning models.

3.3 Random Forest Algorithm

A dependable and effective method for identifying cases of credit card fraud is the Random Forest algorithm. Through the utilization of ensemble learning, this strategy mixes several decision trees to produce a model that is incredibly accurate and dependable. When dealing with imbalanced datasets—that is, datasets where fraudulent transactions are rare. Random Forest excels. Feature selection and bagging approaches are used in tandem to identify subtle trends and abnormalities linked to fraudulent activities. Furthermore, determining what critical elements distinguish genuine from fraudulent transactions is made easier by having the ability to assess the significance of features.

3.4 Support Vector Machine (SVM)

Credit card fraud detection, critical data point identification, and overfitting reduction are all possible with Support Vector Machines (SVM). The system has the ability to process large amounts of data and spot complex fraudulent activity as well as certain attributes. Because of its flexibility, SVM is a trusted choice for banks wishing to increase security and decrease the effect of fraud. The algorithm is flexible and trustworthy.

3.5 K-Nearest Neighbours (KNN)

One reliable and efficient way for identifying instances of credit card fraud is the Random Forest algorithm. By employing ensemble learning, it combines many decision trees to create a model that is very accurate and reliable. Random Forest is highly proficient in handling imbalanced datasets, namely those in which fraudulent transactions are in the minority. The Random Forest algorithm is a highly efficient method for detecting patterns of fraud and anomalies. It accomplishes this by employing bagging and feature selection techniques. It facilitates in distinguishing between genuine and fraudulent transactions by weighing the value of various factors. Since it reduces the possibility of overfitting and provides financial institutions with a sure way to secure clients' cash.

3.6 XG-Boost

Extreme Gradient Boosting (XG-Boost), as widely believed, is an efficacious algorithm in the realm of credit card fraud detection. XGBoost is well known for its outstanding performance. When working with imbalanced datasets where fraudulent transactions are rare, the method's use of gradient boosting techniques results in a robust ensemble model that performs incredibly well. By means of ongoing improvements to the model's performance, it achieves an impressive degree of accuracy in predicting fraudulent behaviors, which allows it to detect even the smallest irregularities with effectiveness. Additionally, financial institutions may precisely identify the critical attributes that support fraud detection according to XG-Boost's feature importance ratings. This enables banking establishments to guarantee the security of transactions and protect their customers against the constantly evolving threats of credit card theft.

4 Results and Analysis

Within this section, we unveil the results of our experimentation with a range of models utilizing the credit card transaction dataset. Following the pre-processing phase, which encompasses data cleansing, normalization, and standardization, we conducted a division of the comprehensive dataset, into 70% and 30% for training and testing purposes respectively (Figs. 4 and 5).

The training dataset forms the bedrock for constructing models, whereas the testing dataset plays a vital role in assessing the effectiveness and performance of these models. An essential tool for assessing the effectiveness of models used to forecast credit card fraud is the Receiver Operating Characteristic (ROC) curve. It shows how, at different categorization thresholds, sensitivity and specificity relate to one another. Achieving a balance between accurately identifying fraudulent transactions (true positives) and minimising the misclassification of legal transactions as fraudulent (false positives) is crucial in credit card fraud detection. This trade-off is visually represented by the ROC curve, which helps analysts choose the best threshold depending on the particular needs of a financial organisation. The top-left corner of an ideal ROC curve indicates high sensitivity and low false positive rates.

Calculating overall performance and detection rate typically depends on the context and the specific metrics you want to use. However, here are common metrics used in many scenarios:

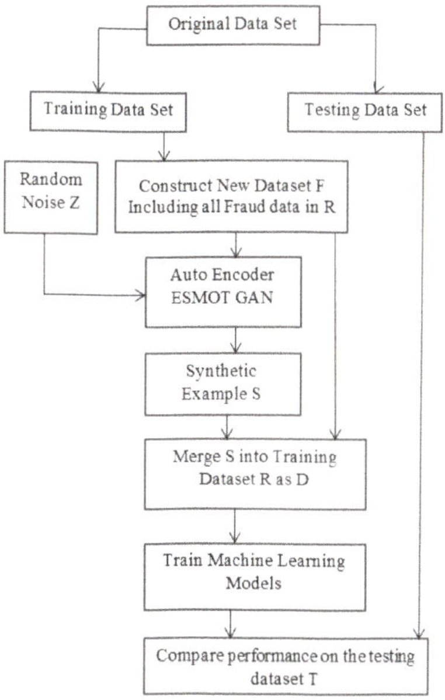

Fig. 4. Credit Card Fraud Detection Flowchart.

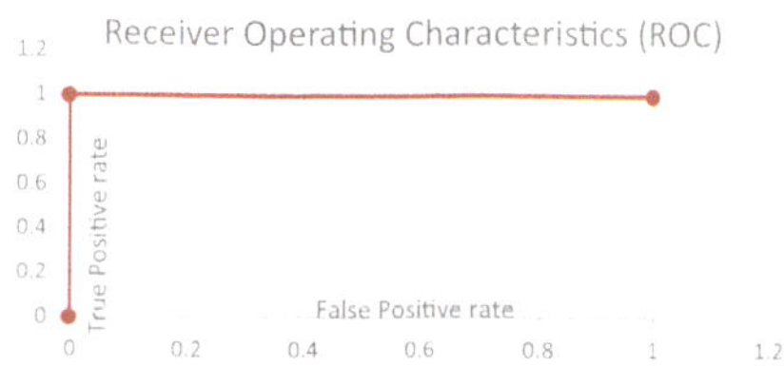

Fig. 5. ROC Curve

4.1 Overall Performance

Overall performance serves as a comprehensive measure of a system or model's effectiveness, often involving a combination of various metrics. A common approach to assess overall performance is by utilizing the F1-score, which calculates the harmonic mean of precision and recall, expressed as

$$F1\text{-}Score = \frac{2 * (Precision * Recall)}{Precision + Recall}$$

Alternatively, accuracy, defined as the ratio of correct predictions to total predictions, can be employed (Fig. 6):

$$Accuracy = \frac{True\ Positives + True\ Negatives}{True\ Positives + False\ Positives + True\ Negatives + False\ Negatives}$$

The performance comparison of various base models coupled with different data augmentation techniques, as measured by the F-measure, highlights the significant impact of these techniques on enhancing credit card fraud detection. Notably, the AE-ESMOTE GAN technique consistently outperforms the others, achieving remarkably high F-measure scores across all base models, reaching almost perfect scores of 99.96 with Random Forest and 99.75 with SVM (Table 1 and Fig. 7).

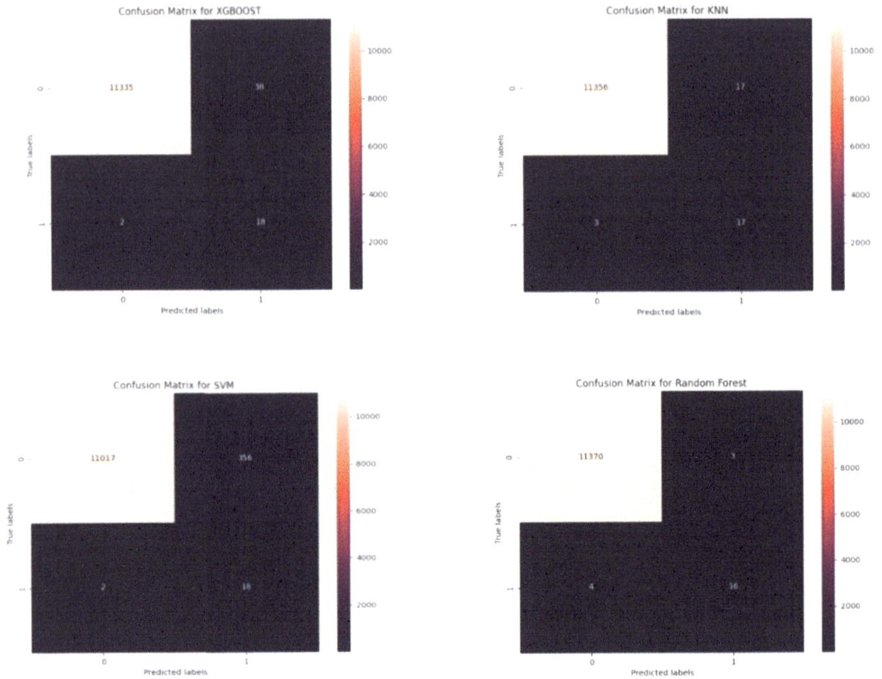

Fig. 6. Confusion Matrix for proposed algorithms

Table 1. Comparitive Analysis in terms of F1 Measure

Base Model	Data Augumentation Techniques						
	Un Balanced	Over Sampling	SMOTE	GAN	SMOTE GAN	ESMOTE GAN	AE-ESMOTE GAN
Random Forest	72.43	89.46	90.43	88.12	91.38	92.31	99.96
SVM	78.71	79.27	89.47	65.18	85.45	87.3	99.75
KNN	80.15	80.48	85.71	81.33	85.71	85.25	99.91
XG-BOOST	83.15	90.14	91.53	87.18	91.53	92.44	97.47

This suggests that the Autoencoder-enhanced Ensemble Synthesized Minority Oversampling with GANs (AE-ESMOTEGAN) is a powerful approach for effectively addressing the class imbalance issue in detecting the credit card fraud. It is evident that the choice of data augmentation technique plays a pivotal role in the performance of the base models, with SMOTE GAN, ESMOTE GAN, and AE-ESMOTE GAN consistently outperforming other methods. This comparison underscores the significance of employing the right data augmentation strategy to optimize the F-measure and, consequently, the accuracy and reliability of credit card fraud detection systems.

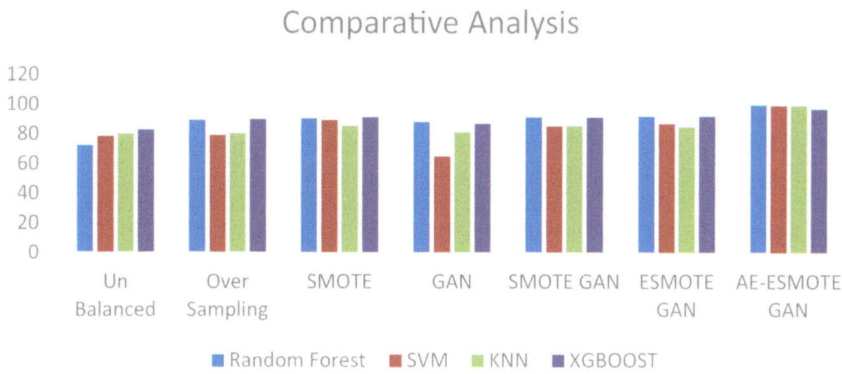

Fig. 7. Comparative Analysis in terms of F1 Measure

5 Conclusion

To conclude, our novel approach for detecting the Credit Card Fraud (AE-ESMOTE-GAN) offers a solid and comprehensive response to the on-going problem of credit card fraud detection by slickly integrating Autoencoders for feature extraction and Ensemble Synthesised Minority Oversampling Techniques with GANs, especially in the context of highly imbalanced datasets dominated by legitimate transactions. By enhancing feature representation quality and tackling class imbalance through synthetic sample generation, our method effectively reduces the risks of over-fitting and significantly improves the accuracy of fraud detection. The evaluation of various classifiers highlights the exceptional performance of our proposed RF algorithm, affirming its potential as a cutting-edge tool for real-time fraud detection. As financial security continues to evolve, our approach is not only innovative but also essential, promising to strengthen security measures and install trust in the dynamic landscape of digital transactions. It provides a promising avenue for financial institutions to overcome the risk of frauds, safeguarding the interests of both organizations and cardholders in an increasingly interconnected and vulnerable financial ecosystem.

Future work should focus on the scalability and adaptability of our algorithm to handle large-scale credit card transaction datasets. Additionally, exploring the integration of more advanced deep learning architectures and real-time monitoring capabilities can enhance the algorithm's efficiency in detecting emerging fraud patterns. Continuous research into improving the interpretability of model decisions and optimizing computational efficiency will further solidify its applicability in the dynamic domain of financial security.

References

1. Ghaleb, F.A., Saeed, F., Al-Sarem, M., Qasem, S.N., Al-Hadhrami, T.: Ensemble synthesized minority oversampling- based generative adversarial networks and random forest algorithm

for credit card fraud detection. IEEE Access **11**, 89694–89710 (2023). https://doi.org/10.1109/access.2023.3306621
2. Dandotia, S., Tiwari, S.K.: Detection of credit card fraud transactions using machine learning algorithms techniques with data driven approaches: a comparative study. Int. J. Innov. Res. Growth **10**(11), 101 (2021). https://doi.org/10.26671/ijirg.2021.11.10.101
3. Vishal Kumar, V.: Credit card fraud detection using machine learning. Comput. Intell. Mach. Learn. **4**(1), 39 (2023). https://doi.org/10.36647/ciml/04.01.a009
4. Ileberi, E., Sun, Y., Wang, Z.: A machine learning based credit card fraud detection using the GA algorithm for feature selection. J. Big Data **9**, 24 (2022). https://doi.org/10.1186/s40537-022-00573-8
5. Singh, S., Chopra, V.: Hybrid machine learning algorithm for credit card fraud detection. Int. Res. J. Modern. Eng. Technol. Sci. **4**, 134 (2022). https://doi.org/10.56726/irjmets29605
6. Samy, N., Mohamed, S.M.: Credit card fraud detection using machine learning techniques. Future Comput. Inform. J. **7**(1), 13–31 (2022). https://doi.org/10.54623/fue.fcij.7.1.2
7. Mohammed, N.H., Maram, S.C.R.: Fraud detection of credit card using logistic regression. SSRN Electron. J. (2022). https://doi.org/10.2139/ssrn.4135514
8. Marabad, S.: Credit card fraud detection using machine learning. Asian J. Converg. Technol. **7**(2), 121–127 (2021). https://doi.org/10.33130/ajct.2021v07i02.023
9. Banu, M., Kavitha, G.: Credit card fraud detection using machine learning algorithms. Int. J. Res. Appl. Sci. Eng. Technol. **10**(10), 1018–1023 (2022). https://doi.org/10.22214/ijraset.2022.47127
10. Bourchouq, A., Peiliang, W.: Design of a model in machine learning for credit card fraud detection. Comput. Eng. Intell. Syst. **13**, 56 (2022). https://doi.org/10.7176/ceis/13-2-06
11. Naik, H., Kanikar, P.: Credit card fraud detection based on machine learning algorithms. Int. J. Comput. Appl.Comput. Appl. **182**(44), 8–12 (2019). https://doi.org/10.5120/ijca2019918521
12. Voican, O.: Credit card fraud detection using deep learning techniques. Inf. Econ. **25**(1\2021), 70–85 (2021). https://doi.org/10.24818/issn14531305/25.1.2021.06
13. Dal Pozzolo, A., Boracchi, G., Caelen, O., Alippi, C., Bontempi, G.: Credit card fraud detection: a realistic modeling and a novel learning strategy. IEEE Trans. Neural Netw. Learn. Syst. **29**(8), 3784–3797 (2018). https://doi.org/10.1109/tnnls.2017.2736643
14. Bodepudi, H.: Credit card fraud detection using unsupervised machine learning algorithms. Int. J. Comput. Trends Technol. **69**(8), 1–3 (2021). https://doi.org/10.14445/22312803/ijctt-v69i8p101

Revolutionizing Agriculture: A Mobile App for Rapid Plant Disease Prediction and Sustainable Food Security

Pasupuleti Sai Kiran[1,2], B. Tirapathi Reddy[1,2], T. Dinesh[1,2](✉), V. Sri Harsha[1,2], S. Harini[1,2], and S. K. Noor Mohammad[1,2]

[1] Prasad V. Potluri, Siddhartha Institue of Technology, Chalasani Nagar, Kanuru, Vijayawada, India
{psaikiran,tirapathireddyb}@pvpsiddhartha.ac.in, dineshnarayan2003@gmail.com

[2] Springer Heidelberg, Tiergartenstr. 17, 69121 Heidelberg, Germany

Abstract. In order to maintain the health of crops and the quality of the yield they offer, it is very important to diagnose plant illnesses in a timely manner and conduct quality assessments of them. A number of diseases, including bacterial spots, late blight, and leaf mold, amongst others, have the potential to have a detrimental influence on plant production. The primary goal of this research is to develop a mobile application that uses a deep learning model and can predict leaf diseases. In this investigation, a convolutional neural network, more commonly referred to as a CNN, is used on a dataset that includes images that the researchers themselves collected as well as photographs of diverse plant leaves that were acquired from the collection that is stored at Plant Village. In other words, the pictures in the dataset come from two different sources. Using the CNN model, accurate predictions have been produced in excess of 94% of the time.

Keywords: Deep Learning · Android Application · CNN

1 Introduction

Plant diseases are a decline from a plant's original state that disrupts and changes the plant's essential functions. Farmers and agronomists deal with plant diseases on a regular basis. Numerous factors, such as the variety of nearby crops, the presence of pathogens, and environmental conditions, can contribute to plant diseases. Most diseases initially show symptoms that can be identified by looking for changes in the physical attributes of the plant, such as alterations in size, shape, or color.

To tackle these challenges, we rely on our in-depth understanding of the various plant diseases that we obtain through the analysis of leaf symptoms that are documented in photographs. Farmers may suffer large yearly financial losses as a result of various diseases that can affect vegetable crops like potatoes, tomatoes, and peppers. For example, there are two types of blight: late blight that is brought on by a particular

V. E. Balas et al. (Eds.): IEQIP 2023, LNNS 1029, pp. 17–26, 2024.
https://doi.org/10.1007/978-3-031-61929-8_3

bacterium and early blight that is brought on by a fungus. Farmers can lessen losses and costs by detecting these diseases early and implementing efficient treatments. In the next 25 years, it is expected that there will be more people on the planet than 9 billion, which will increase demand for food production by 70%. Crop diseases, however, are a serious problem, especially in nations that are agrarian. Because potatoes are so widely consumed, diseases affecting them are especially concerning. However, illnesses in tomatoes and peppers are equally concerning. Agricultural experts are investigating the possibility of using deep learning to proactively fight potato crop diseases with the age-old maxim "prevention is better than cure" in mind. Machine learning techniques have demonstrated potential in extracting data from real-time image processing, owing to notable advancements in computing.

There is no denying the significance of deep learning, which is becoming more and more popular, improving people's everyday lives and showing promise for the future. Plant diseases affect the quality and yield of crops, including leaves, fruits, stems, and roots, which results in a global shortage of vegetables consumed. In actuality, crop diseases cause a 16% annual loss in crop productivity. Deep learning is widely used in smart farming to incorporate cutting-edge techniques, instruments, and algorithms into the agricultural environment. Agricultural problems like feature extraction, transformation, pattern recognition, and image classification can be solved, especially with machine learning. Research efforts in this area are focused on creating a Convolutional Neural Network (CNN)-based method to identify disease types and detect crop infections. The ultimate objective is to develop a system that can provide results in real-time and possibly drastically lower crop yield losses.

In remote regions where access to agricultural specialists is restricted, farmers frequently depend on their own expertise and understanding to identify crop illnesses. Specialists can occasionally diagnose these illnesses, but the procedure is frequently costly, labor-intensive, and time-consuming, making it unfeasible for widespread application in developing countries. On the other hand, rapid disease detection through the use of machine learning and deep learning techniques can significantly lower crop damage. At the forefront of these technological advances are CNNs, which are being used to develop intelligent systems capable of diagnosing plant diseases from pictures of the leaves. These systems provide accurate diagnoses of the precise diseases affecting the plants by closely examining visual cues such as color variations, spots, lesions, and patterns.

2 Related Work/Literature Survey

Many publications in the field of image-based plant disease identification typically follow a basic procedure that includes background removal and the isolation of diseased plant lesion tissue as first preprocessing steps. To enable a more thorough examination of the plant disease, discriminative features are then extracted, leading to the creation of a classification model. The technique used by Al-Hiary et al. [1] involved identifying green-colored pixels and masking them according to predetermined threshold values; as a result, accuracy levels between 83% and 94% were attained.

In order to maintain high performance standards, Mitali [2] optimized the CNN architecture by finding the perfect ratio between layers and trainable parameters. In order to find the best model parameters for testing, this process involved fine-tuning the model [2]. By doing away with the need for manual feature engineering and preprocessing, their feed-forward network successfully reduced computational complexity and saved time. Pooling and convolution layers in the network allowed for automatic feature extraction.

The CNN model proposed by Shahina [3] has seven convolutional layers (3x3 filters), four pooling layers, two fully connected layers, and uses batch normalization and dropout to address overfitting. It also has RELU activation, consistent padding, a stride of 1, and Adam optimizer optimization. In comparison to a traditional deep neural network, Sanasam Premananda [4] significantly lowered the number of parameters by combining CNN features like parameter sharing and sparsity of connection.

The model developed by Showmick Guha Paul [5] has two phases: the deployment phase, in which the best-performing models are chosen and integrated into an Android application and a local web-based system for the prediction of tomato leaf disease, and the training phase, which involves data collection, preprocessing, and training with multiple models. Laura Falaschetti [6] utilized the OpenMV Cam to enhance result precision [7], the authors employed a neural network that considered both color and shape properties of infected leaves to examine diseases in rice crops [8]. In proposed model involves the smaller no. of parameters than the pre-trained models evaluated in this work.

Research papers [9] encompass a range of experiments conducted for the detection of tomato leaf diseases [10]. The paper introduces a ResNet-9 model designed to detect blight illnesses in potato and tomato crops at an early stage. The model achieves an accuracy of 99.25% and generates saliency maps that may be used by farmers [11]. This paper utilized a convolutional neural network (CNN) structure comprising of three convolutional layers, three max pooling layers, and two fully connected layers [12]. The authors of this paper utilized VGGNet as the recommended model, which is a pre-trained model [13]. In this paper, the authors employed DCNN (Deep Convolutional Neural Network) instead of CNN (Convolutional Neural Network) in order to achieve superior outcomes. The papers [14, 15] explore the application of convolutional neural networks (CNN) for detecting crop diseases. Papers [16, 17] present diverse methods and outcomes of experiments conducted for predicting weeds in corn plants. Papers [18, 19] include models and results related to the prediction of diseases in pepper and potato plants. Paper [20] include models and results related to the prediction of disease in the grapes. Papers [21–23] provide information on various models applicable for plant disease prediction like svm, vgg16, vgg19 etc., while [24] offers a comparison of different pretrained models based on their performance.

Sandra Kumi [25] focuses on automatic disease detection, primarily utilizing Convolutional Neural Networks (CNNs). Four distinct CNN models were trained and assessed, with SSD MobileNet V2 emerging as the most effective model, demonstrating a confidence level of over 80% in disease detection.

3 Proposed System

We used TensorFlow's tf.keras.preprocessing.image.ImageDataGenerator to bolster our image dataset's robustness and diversity, crucial for our deep learning model's performance and generalization.

Our aim was to enhance our image data while preserving its integrity. We standardized pixel values by dividing by 255 for consistency. We introduced diversity through random 20-degree rotations, 20% horizontal and vertical shifts, shearing, and zooming. Horizontal flips added mirror images for dataset enrichment. Additionally, we incorporated a 'fill_mode' parameter set to 'nearest,' which ensured that any gaps resulting from the aforementioned transformations were filled with the nearest available pixel values, preserving the overall quality of the images.

In our research, we present a meticulously crafted deep convolutional neural network (CNN) architecture for image classification tasks, aimed at achieving state-of-the-art performance and robustness. The architecture is intricately designed to effectively capture fine-grained features within input images, enabling the model to make highly accurate predictions.

Our CNN architecture consists of several convolutional blocks, each serving the purpose of progressively extracting and learning crucial features from the input data. The first convolutional block commences with 32 filters, each of size 3×3, followed by batch normalization and ReLU activation functions. This combination enhances the model's capacity to recognize intricate patterns within the data. A subsequent max-pooling layer is employed to reduce the spatial dimensions and retain the most informative features.

Successive convolutional blocks follow a similar pattern, effectively doubling the number of filters while maintaining the essential batch normalization and RELU activation functions. These blocks serve to capture hierarchical features, enabling the network to comprehend both low-level and high-level representations in the data. Further down sample is achieved through additional max-pooling layers, which ensures the processing of increasingly abstract information.

Upon the completion of the final convolutional block, we transition to the flattening of the feature maps into a vector format. Dropout layers are strategically integrated into the network to counteract overfitting, thereby improving its generalization capabilities. The fully connected layers include a dense layer comprising 1024 units with ReLU activation, followed by an additional dropout layer to further enhance robustness. The model's output layer is equipped with a softmax activation function, yielding class probabilities tailored to our specific classification task.

The resulting model represents a powerful image classifier, adept at handling diverse complexities and subtleties within the input data. Its architectural design is carefully calibrated to extract and encapsulate meaningful information, while regularization techniques such as dropout ensure the model's capacity to generalize effectively. Our deep learning model stands as a noteworthy contribution to the field of image classification, offering a robust and highly accurate solution for a wide array of applications.

We used the highly regarded 'adam' optimizer to enable effective learning. The Adam optimizer offers an adaptive learning rate mechanism that effectively optimizes model parameters, combining the advantages of the Adagrad and RMSprop optimizers. This adaptive optimization technique is particularly beneficial in our complex image classification challenge because it dynamically adjusts the learning rates for each parameter. This facilitates the model's rapid convergence and enhances the accuracy of its output.

We selected the 'categorical cross-entropy' as our loss function, which is a key choice for addressing multiclass classification problems. The utilization of this loss function provides the network with a distinct and unambiguous indication to decrease the mistake in classification by precisely measuring the disparity between the model's predictions and the true class labels.

In terms of performance evaluation, we have chosen to closely monitor the metric of 'accuracy'. We have implemented dynamic learning rate adjustment to enhance the training procedure of our deep convolutional neural network. Using the 'ReduceLROnPlateau' callback is essential in this method since it optimizes the performance of our model. The objective of this callback is to monitor and record the validation losses that arise throughout the training process. The system operates by tracking variations in the validation loss and adaptively modifying the learning rate to facilitate the model's convergence when a plateau is detected.

In particular, our 'ReduceLROnPlateau' callback is set up to watch the validation loss and lower the learning rate by a factor of 0.5 if it perceives no improvement over a predetermined 'patience' of three training epochs. This decrease in learning rate guarantees that the model learns efficiently even in the presence of complex or slowly changing data patterns, helping to overcome any convergence difficulties. In order to keep the learning rate from dropping too much, we have also set a minimum learning rate ('min_lr') of 1e-6.

To deploy our deep learning model as quickly and effectively as possible, we used TensorFlow Lite post-training quantization techniques. The model's size is decreased and its inference speed is increased thanks to this optimization step, which makes it especially appropriate for devices with limited resources.

The procedure involved converting our Keras model into a quantized TensorFlow Lite model using the TFLiteConverter. Importantly, we used the 'DEFAULT' optimization level, which optimizes the model for better inference performance and smaller size.

The TensorFlow Lite model that resulted from the quantization process was saved to a file so that it could be deployed on different edge devices. This optimization is a crucial stage in our research because it improves the usefulness and effectiveness of our plant disease prediction system, especially when it is implemented on embedded and mobile devices. This will lead to more effective and accessible agricultural practices (Fig. 1).

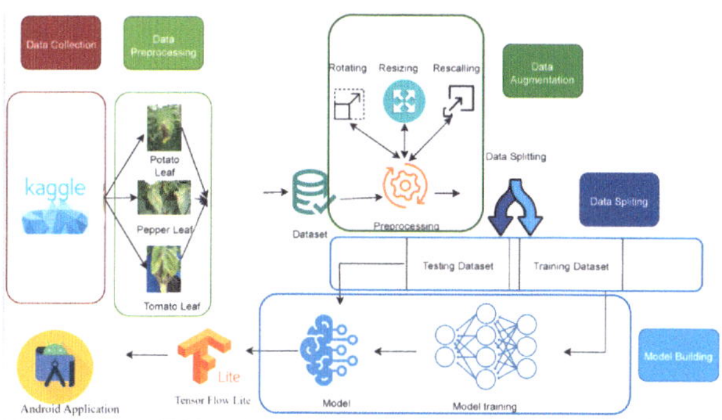

Fig. 1. Plant disease detection using Android Application Architecture Diagram

4 Technologies Used

4.1 Google Colab

In our research, Google Colab proved to be an essential cloud-based environment, providing free access to GPUs and TPUs. For our work, it eliminated hardware limitations and the need for local installations, providing an effective platform. Our workflow was greatly streamlined and setup time was greatly decreased by its pre-installed libraries and support for real-time collaboration among multiple users. Because of this, Google Colab is the best option for projects involving a lot of data and teamwork.

4.2 Android Studio

As the official integrated development environment (IDE) for Android app development, Android Studio played a pivotal role in our research. This powerful and feature-rich IDE enabled us to create, test, and debug Android applications efficiently. With its user-friendly interface and a wide array of tools, Android Studio simplified our app development process. Notable features such as code completion, intelligent code analysis, and an intuitive drag-and-drop interface designer contributed to its effectiveness.

4.3 Data Set

The Plant Village Dataset, consisting of 20,000 images of various plants, was obtained from Kaggle and 80% of data is used for the training purpose and remaining 20% of images i.e. around 4,000 images are used for the testing purpose.

4.4 Modules

- **Data Collection**: The dataset is categorized into 'Train' and 'Test,' [26] further segmented into various plant classes such as potato, tomato, corn, and pepper.
- **Training Module**: Within this module, our model is meticulously trained to recognize leaf diseases based on user-input data.
- **Testing Module:** This module serves to evaluate the model's predictive capabilities, offering disease predictions for user-provided input.
- **Accuracy:** To gauge the performance of our system, we conduct multiple training epochs and calculate accuracy using a predefined formula.

Together, these modules make up the core of our plant disease prediction system, which helps to progress agriculture and food security by facilitating effective data processing, model training, and accurate disease diagnosis.

5 Results

The plant disease prediction system, utilizing a convolutional neural network (CNN), has demonstrated exceptional performance in accurately diagnosing leaf diseases. After undergoing thorough testing and review, our model demonstrated a remarkable accuracy rate of 94% [27]. The exceptional level of accuracy demonstrates the strength and effectiveness of our deep learning model in categorizing plant diseases (Fig. 2).

Fig. 2. Sample Images From the dataset

Moreover, by utilizing TensorFlow Lite, our system has effectively been implemented on end devices, specifically on platforms based on Android. By deploying our plant disease prediction technology to mobile and embedded devices with limited resources, we guarantee that farmers and agricultural enthusiasts may use it while working in the field. The high efficiency of TensorFlow Lite in deploying models on such devices highlights its aptness for edge computing and mobile applications, hence improving the practicality and accessibility of our service (Figs. 3 and 4).

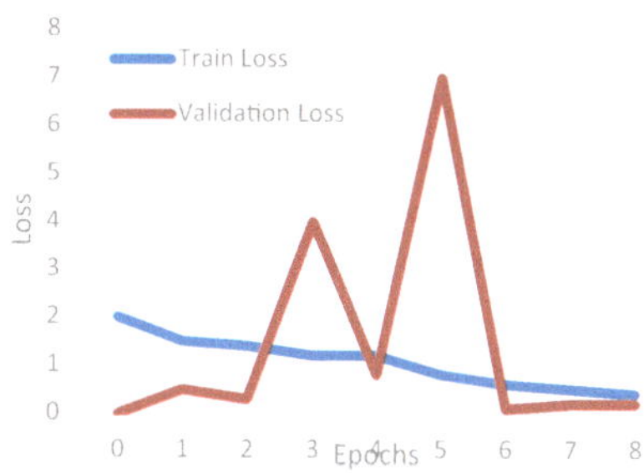

Fig. 3. Training and Validation Loss Graph

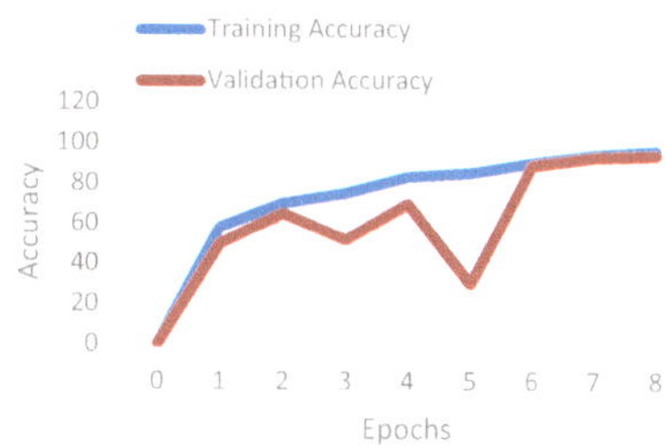

Fig. 4. Accuracy Graph during training and validation

These findings showcase the efficacy of our plant disease prediction system in the early detection and management of illnesses, thereby enhancing food security and advocating for sustainable agricultural practices (Figs. 5 and 6).

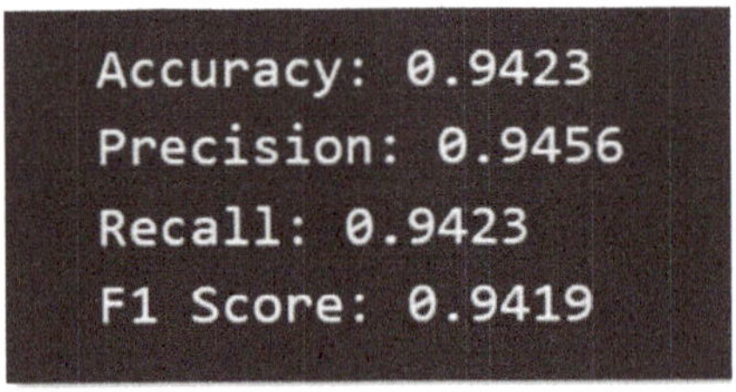

Fig. 5. Performance metrics of the model

Fig. 6. Mobile app detecting that plant has disease

6 Scope and Future Use

The potential uses of deep learning and an Android app for diagnosing leaf diseases are broad and promising. For the purpose of safeguarding crop yields and food production, farmers and other agriculturalists might employ it as a valuable instrument to promptly identify and address plant diseases.

There are numerous potential applications in the future. Through the integration of this technology into intelligent farming systems, it becomes possible to monitor vast agricultural areas for illnesses in real-time. Early outbreak detection and disease mapping can enhance the efficacy of pest and disease control approaches. Furthermore, with the expansion of the dataset, the deep learning models will exhibit enhanced precision and proficiency in recognizing a broader spectrum of diseases and plant species. Moreover, the Android application extends the reach of this technology to a worldwide user population, granting farmers across the globe the opportunity to utilize a tool that has the capability to revolutionize crop safeguarding and enhance food security. Deep learning and Android apps for leaf disease identification hold significant potential for the future of sustainable agriculture and improved crop management.

7 Conclusion

The research introduces a comprehensive framework that utilizes mobile technologies, deep learning, and efficient model implementation to predict plant diseases. Plant diseases unquestionably present a substantial peril to the long-term viability of agriculture and the stability of global food supply. It is imperative to tackle this issue with pragmatic and effective solutions. The advancements we have achieved in this domain are substantial.

We have enabled farmers and agricultural enthusiasts to quickly and accurately diagnose leaf diseases in a range of plant classes by developing a mobile application. This easily comprehensible and economical solution is based on a convolutional neural network (CNN) with an astounding 94% accuracy rate. Early intervention is made possible by the high degree of accuracy, which may help to reduce crop damage and yield losses.

TensorFlow Lite enhances Android smartphone system's agricultural capabilities, making it suitable for limited resource devices, making it an efficient choice for mobile applications.

Sustainable farming practices, food security, and agriculture are all improved by our research. Our ability to predict plant diseases accurately helps with early disease diagnosis, which is essential for minimizing crop losses and guaranteeing the world's food security. This work signals the beginning of a promising era for data-driven and intelligent farming practices by showcasing the potential of mobile technology, artificial intelligence, and effective model deployment to address pressing agricultural challenges.

References

1. Al Hiary, H., Bani Ahmad, S., Reyalat, M., Braik, M., Al Rahamneh, Z.: Fast and accurate detection and classification of plant diseases. Int. J. Comput. Appl. **17**(1), 31–38 (2011). https://doi.org/10.5120/2183-2754
2. Shewale, M.V., Daruwala, R.D.: High performance deep learning architecture for early detection and classification of plant leaf disease. J. Agric. Food Res. **14**, 100675 (2023). https://doi.org/10.1016/j.jafr.2023.100675
3. Anwarul, S., Mohan, M., Agarwal, R.: An unprecedented approach for deep learning assisted web application to diagnose plant disease. Procedia Comput. Sci. **218**, 1444–1453 (2023). https://doi.org/10.1016/j.procs.2023.01.123
4. Singh, S.P., Pritamdas, K., Devi, K.J., Devi, S.D.: Custom convolutional neural network for detection and classification of rice plant diseases. Procedia Comput. Sci. **218**, 2026–2040 (2023). https://doi.org/10.1016/j.procs.2023.01.179
5. Paul, S.G., Al Biswas, A., Saha, A., Zulfiker, M.S., Ritu, N.A., Zahan, I., Rahman, M., Islam, M.A.: A real-time application-based convolutional neural network approach for tomato leaf disease classification. Array **19**, 100313 (2023). https://doi.org/10.1016/j.array.2023.100313
6. Falaschetti, L., Manoni, L., Di Leo, D., Pau, D., Tomaselli, V., Turchetti, C.: A CNN-based image detector for plant leaf diseases classification. HardwareX **12**, e00363 (2022). https://doi.org/10.1016/j.ohx.2022.e00363
7. Orillo, J.W., Dela Cruz, J., Agapito, L., Satimbre, P.J., Valenzuela, I.: Identification of diseases in rice plant (Oryza sativa) using back propagation artificial neural network. In: Proceedings of International Conference on Humanoid Nanotechnology Information Technology of Communication and Control Environmental Management (HNICEM), pp. 1–6 (2014)
8. Vishnoi, V.K., Kumar, K., Kumar, B., Mohan, S., Khan, A.A.: Detection of apple plant diseases using leaf images through convolutional neural network. IEEE Access **11**, 6594–6609 (2023). https://doi.org/10.1109/access.2022.3232917
9. Gontijo da Cunha, V.A., Hariharan, J., Ampatzidis, Y., Roberts, P.D.: Early detection of tomato bacterial spot disease in transplant tomato seedlings utilising remote sensing and artificial intelligence. Biosys. Eng. **234**, 172–186 (2023)
10. Anim-Ayeko, A.O., Schillaci, C., Lipani, A.: Automatic blight disease detection in potato (Solanum tuberosum L.) and tomato (Solanum lycopersicum, L. 1753) plants using deep learning. Smart Agric. Technol. **4**, 100178 (2023)
11. Agarwal, M., Singh, A., Arjaria, S., Sinha, A., Gupta, S.: ToLeD: tomato leaf disease detection using convolution neural network. Procedia Comput. Sci. **167**, 293–301 (2020). https://doi.org/10.1016/j.procs.2020.03.225
12. Sanida, T., Sideris, A., Sanida, M.V., Dasygenis, M.: Tomato leaf disease identification via two–stage transfer learning approach. Smart Agric. Technol. **5**, 100275 (2023). https://doi.org/10.1016/j.atech.2023.100275

13. Hossain, M.I., Jahan, S., Al Asif, M.R., Samsuddoha, M., Ahmed, K.: Detecting tomato leaf diseases by image processing through deep convolutional neural networks. Smart Agric. Technol. **5**, 100301 (2023)
14. Mishra, S., Sachan, R., Rajpal, D.: Deep convolutional neural network based detection system for real-time corn plant disease recognition. Procedia Comput. Sci. **167**, 2003–2010 (2020). https://doi.org/10.1016/j.procs.2020.03.236
15. Wessner, R.N., Frozza, R., da Duarte Silva Bagatini, D., Molz, R.F.: Recognition of weeds in corn crops: System with convolutional neural networks. J. Agric. Food Res. **14**, 100669 (2023). https://doi.org/10.1016/j.jafr.2023.100669
16. Suárez, P.L., Velesaca, H.O., Carpio, D., Sappa, A.D.: Corn kernel classification from few training samples. Artif. Intell. Agric. **9**, 89–99 (2023). https://doi.org/10.1016/j.aiia.2023.08.006
17. Prakash K.B.: Content extraction studies using total distance algorithm. In: Proceedings of the 2016 2nd International Conference on Applied and Theoretical Computing and Communication Technology, iCATccT 2016, Art. No. 7912085, pp. 673–679 (2017)
18. Subeesh, A., et al.: Deep convolutional neural network models for weed detection in polyhouse grown bell peppers. Artif. Intell. Agric. **6**, 47–54 (2022)
19. Prakash, K.B.: Mining issues in traditional Indian web documents. Indian J. Sci. Technol. **8**(32), 1–11 (2015). https://doi.org/10.17485/ijst/2015/v8i1/77056
20. Praveen, S.P., Nakka, R., Chokka, A., Thatha, V.N., Vellela, S.S., Sirisha, U.: A novel classification approach for grape leaf disease detection based on different attention deep learning techniques. Int. J. Adv. Comput. Sci. Appl. **14**(6), 1199 (2023). https://doi.org/10.14569/ijacsa.2023.01406128
21. Islam, M.M., Adil, M.A.A., Talukder, M.A., Ahamed, M.K.U., Uddin, M.A., Hasan, M.K., Sharmin, S., Rahman, M.M., Debnath, S.K.: DeepCrop: deep learning-based crop disease prediction with web application. J. Agric. Food Res. **14**, 100764 (2023). https://doi.org/10.1016/j.jafr.2023.100764
22. KumarSahu, S., Pandey, M.: An optimal hybrid multiclass SVM for plant leaf disease detection using spatial fuzzy C-means model. Expert Syst. Appl. **214**, 118989 (2023). https://doi.org/10.1016/j.eswa.2022.118989
23. Madhuri, A., Jyothi, V.E., Phani Praveen, S., Altaee, M., Abdullah, I.N.: Granulation-based data fusion approach for a critical thinking worldview information processing. J. Intell. Syst. Internet Things **9**(1), 49–68 (2023). https://doi.org/10.54216/JISIoT.090104
24. Jackulin, C., Murugavalli, S.: A comprehensive review on detection of plant disease using machine learning and deep learning approaches. Meas. Sens. **24**, 100441 (2022). https://doi.org/10.1016/j.measen.2022.100441
25. Kumi, S., Kelly, D., Woodstuff, J., Lomotey, R.K., Orji, R., Deters, R.: Cocoa companion: deep learning-based smartphone application for cocoa disease detection. Procedia Comput. Sci, **203**, 87–94 (2022). https://doi.org/10.1016/j.procs.2022.07.013
26. Prakash, K.B.: Quantum meta-heuristics and applications. In: Prakash, K.B., Kanagachidambaresan, G.R., Srikanth, V., Vamsidhar, E. (eds.) Cognitive Engineering for Next Generation Computing: A Practical Analytical Approach, pp. 265–297. Wiley, Hoboken (2021). https://doi.org/10.1002/9781119711308.ch10
27. Prakash, K.B.: Data Science Handbook: A Practical Approach, pp. 1–453. Wiley, Hoboken (2022)

EMG Based Human Machine Integration for IoT Based Instruments

V. Rajesh[1], V. Sri Sravan[2], and N. Nanda Prakash[1](✉)

[1] Department of ECE, Koneru Lakshmaiah Education Foundation, Vaddeswaram, Guntur 522302, A.P., India
nandaprakashnelaturi@gmail.com

[2] Department of CSE, Koneru Lakshmaiah Education Foundation, Vaddeswaram, Guntur 522302, A.P., India

Abstract. Introducing the application of hand gestures in human-computer interaction, this study presents a novel approach that employs artificial neural networks (ANNs) to analyze muscle activity and distinguish hand movements from electromyograms (EMGs). The emphasis on creating a user-friendly and stress-free environment is reflected in the limitation of EMG detection to three specific muscle groups in the arm. The analysis of EMG signals yields parameters related to the temporal patterns of muscle contractions, each uniquely associated with a specific gesture. An experimental study was conducted to assess the system's ability to accurately recognize six distinct arm movements. The results demonstrate that the algorithm achieves an impressive 98% accuracy rate in identifying all six motions. One notable advantage of this system is its ease of training and seamless real-time execution once the training process is completed.

1 Introduction

Traditional computer interfaces, such as keyboards and mice, rely heavily on the user's physical movements and tactile interactions. The wrist, palm, fingers, and upper limbs are all involved in these precise and deliberate physical interactions. However, there are scenarios in which individuals with physical impairments, such as those resulting from conditions like spinal cord injuries (SCI), paralysis, or amputations, face significant challenges in using these interfaces. Ongoing efforts are dedicated to addressing the growing accessibility barriers faced by individuals with disabilities when it comes to using computer interfaces. Modern technology has provided solutions for individuals with impairments to access the IT environment without relying on devices that necessitate direct physical contact, including devices operated by mouth or forehead swabs. These devices face issues related to handling complexities and a lack of precision. Given that biosignals can convey information about body movements more swiftly than alternative mediums like video sequences, they have emerged as a novel focal point for addressing the challenges associated with extracting user intent. Various biosignal-based human-machine interfaces (HCI) have proven successful in this endeavor, leading to the development of interactive interfaces that empower users in the face of these obstacles. While biosignal-based human-computer interaction (HCI) has shown success, there's a

V. E. Balas et al. (Eds.): IEQIP 2023, LNNS 1029, pp. 27–34, 2024.
https://doi.org/10.1007/978-3-031-61929-8_4

lack of widely accepted performance benchmarks within academic circles. Researchers tend to design interfaces using similar foundational methods; however, the standards and criteria for assessment vary significantly, making it difficult to compare the effectiveness of different interfaces. Increasingly, experts in the fields of robotics and HCI emphasize the importance of developing more user-friendly interfaces to humanize the interaction between humans and machines. Achieving this objective necessitates improvements in both software and hardware systems. Users have growing expectations for the technical specifications of their work environments. They desire the ability to interact naturally, using gestures and other innate expressions. Consequently, the integration of gestures and sign language into advanced interface design has gained significant importance in recent times. One of the driving factors behind this surge of interest in gesture-based interfaces is the broad spectrum of activities and applications that can be enhanced and simplified through natural manipulation and processing. Presently, examples of natural and intuitive interfaces include arm/hand pads and data gloves. However, these methods require users to employ bulky equipment that may limit their natural movements. Additionally, data gloves may not be suitable for precise object manipulation. Alternative approaches involve information processing systems for audio, video, and electromyography (EMG) (Fig. 1).

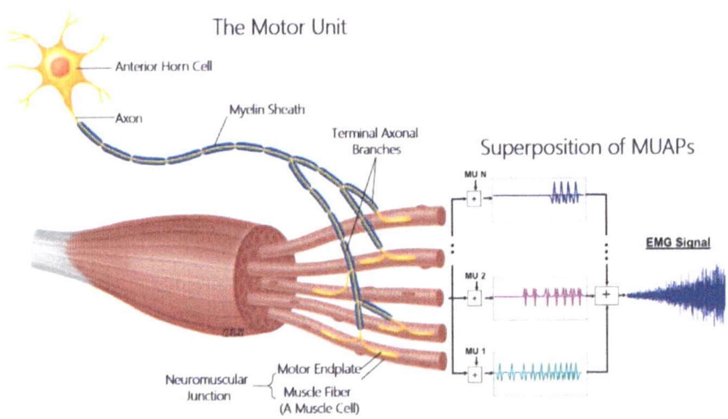

Fig. 1. The collective action potentials from multiple motor units, detectable through electrodes, overlay to create the electromyography (EMG) signal.

Voice recognition technology, currently available, lacks reliability, particularly in noisy environments, with multiple speakers, and in regulated outdoor settings. In contrast, video data capture doesn't restrict the user in any way, but even with multiple cameras, obtaining comprehensive movement information can be exceedingly challenging, often leading researchers to limit the application's scope. Furthermore, it's vital to measure the power, pressure, and effort applied by the user's hand while interacting with machines, and these aspects cannot be accurately determined through visual information alone. The application of EMG signals for trajectory tracking has been a successful part of medical technology for over three decades. Nevertheless, EMG-based control systems have recently taken a new direction, as studies have proposed using EMG signals as a means of communication with machines. However, individuals with progressive neuromuscular issues may encounter significant challenges. The primary barrier arises from muscle weakening and limited range of motion, which is further exacerbated by reduced capabilities that may render dynamic proportional control systems impractical. Typically, EMG assistive devices are considered as a last resort when patients reach this stage of their condition. This approach challenges the prevailing belief in the field of rehabilitation technology, where EMG activity is commonly utilized for computer control and, more recently, for prosthetic device control. However, its application in automated equipment, such as coarse movement, has not gained much attention.

2 Methodology

2.1 Feature Selection

To operate, it requires four distinct movements that are both simple to execute and yield highly distinguishable patterns in the EMG signal. Prior to and after each gesture, there's a specific hand positioning known as the starting position, which should ideally produce minimal signal amplitude. Following extensive testing of over 20 different gestures, the four control gestures that best met the criteria were selected, as outlined in Table 1. The first gesture involves a brief hand stretch (Gesture I), serving as the initial gesture's entry point. To initiate the second gesture, which is the hand grip (Gesture 2), a swift and gentle pressure is applied to the first. The third gesture is wrist extension (Gesture 3), which involves an outward movement of the wrist. The fourth and final gesture chosen is thumb flexion (Gesture 4), characterized by the thumb bending inward [1–3]. Our research findings indicate that, except for minor variations in overall amplitude, both hands produce very similar EMG waveforms. As a result, determining which hand is engaged in the process does not present a challenge. The titles and descriptions of the motions are oriented towards the right hand because it is the preferred choice for most users; however, these titles and descriptions can be adapted for left-handed users to mirror the same gestures.

2.2 System Structure

This section provides an in-depth insight into personalized care within the system. It describes the sequence of actions, starting from capturing a gesture to selecting a Human-Computer Interaction (HCI) option [4, 5]. The process begins with electrodes connected to a sensor, recording the EMG signal generated by the muscles involved in arm movements. The recorded signal undergoes preprocessing to prepare it for further analysis. Subsequently, incoming patterns that correspond to the gestures within the signal are matched. Key attributes of each model are initially retrieved to distinguish them. The classification of the sequence of movements is achieved using the feature vector generated. Finally, the microcontroller executes the control command transmitted through the gesture to operate peripheral devices.

2.3 Signal Acquisition

Our EMG sensor is a surface unit designed to capture EMG signals within the active frequency range of 20 to 500 Hz, with a maximum voltage capacity of 6000 μV. To record the EMG signal, we employed only two separate Ag/AgCl electrodes attached to the user's skin. These electrodes were affixed to the abductor pollicis longus muscle, which is connected to the sensor. Typically, each electrode pair is dedicated to the investigation of a specific muscle, with the aim of avoiding crosstalk or undesirable signal interference from neighboring muscles. As we had access to just one sensor channel, we used one electrode pair to analyze multiple muscles simultaneously (Fig. 2).

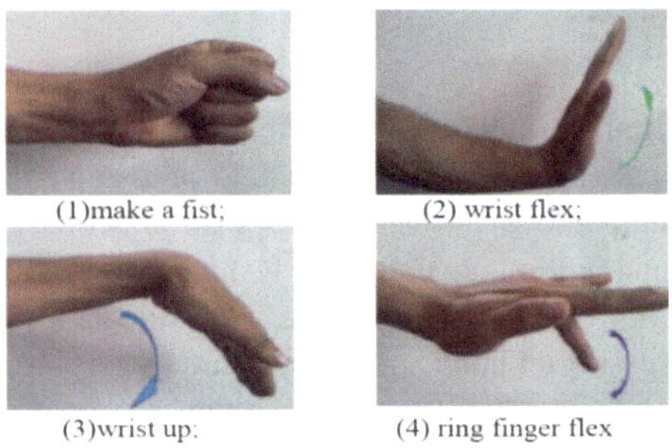

Fig. 2. Utilization of the EMG benchtop system through four distinct hand gestures.

2.4 MUAPS Using Wavelets

The electrodes attached to muscles capture a mix of background noise, artifacts, and single motor unit action potentials (MUAPs). Surface MUAPs' fundamental shapes can be adequately described by only a few wavelet functions [9–11]. Two clinically significant aspects of the EMG signal are the MUAP waveform and the number of activated motor units. Clinical electromyography (EMG) quantitative analysis is highly desirable, and various methods have been employed for quantitative analysis in both the time and frequency domains with the advent of computerized EMG devices [12–14]. The wavelet transform represents time-frequency in two dimensions. It employs both a short high-frequency window and a long low-frequency window. In the case of the db4 wavelet, at the highest frequency scales, the wavelet transform coefficients offer an excellent temporal resolution of just four signal samples. This enables effective tracking of major MUAP peak transients, a level of temporal resolution not achievable with STFT [15–18]. In our research, we decomposed the signals using four layers of db4.

2.5 Feature Extraction

A performance metric was employed to identify and adapt distinctive characteristics from each combined model for gesture utilization. In this process, several statistical features like RMS, entropy, and standard deviation were omitted. In pattern recognition, Artificial Neural Networks (ANNs) have become a significant technique in Human-Computer Interaction (HCI) research [19, 20]. One of the advantages of using ANNs is that they function as black box models, reducing the need for an in-depth understanding of the system. The network, often referred to as the black box, is developed by processing numerous samples through it. Once the training process is successfully completed, the network remains intact and is presented with new, previously unseen input data. Subsequently, the network processes the data to generate the required outputs. All participants were required to perform six distinct wrist movements in each session during the training phase. Filtered EMG data was then collected [21, 22]. These six sets of wrist motions, along with the corresponding network responses outlined in Table 1, were employed to train the network. A backpropagation algorithm with a boosting technique was used for optimization.

2.6 Experimental Setting

The experiment was conducted again, this time involving a total of 20 participants. Initially, we collected personal data from each participant, including their age (average: 21.85 years), gender (nine females and eleven males), weight (average: 52.15 kg), and hand dominance (all 20 were right-handed, none left-handed). The experiment comprised two stages. In the first phase, ten subjects participated, and we recorded six hand position gestures for each subject, resulting in a total of 60 gestures. For each gesture, we measured three feature values, calculated their averages, and subsequently used these average features to evaluate the features for another set of ten subjects. The remaining ten participants served as the test sample.

3 Results and Observations

The backpropagation neural network was trained using four distinct hand motions, which were repeated in five separate experiment sets. When the ANN was trained with these normalized values from a single participant's data, the results indicated a high degree of convergence. The outcomes of the ANN testing, employing the weight matrix generated from the training data, are presented in Table 1. Accuracy was assessed by calculating the percentage of correctly identified data points in relation to the total number of data points. In each experiment, the categorization accuracy reached 97.5% (Figs. 3 and 4).

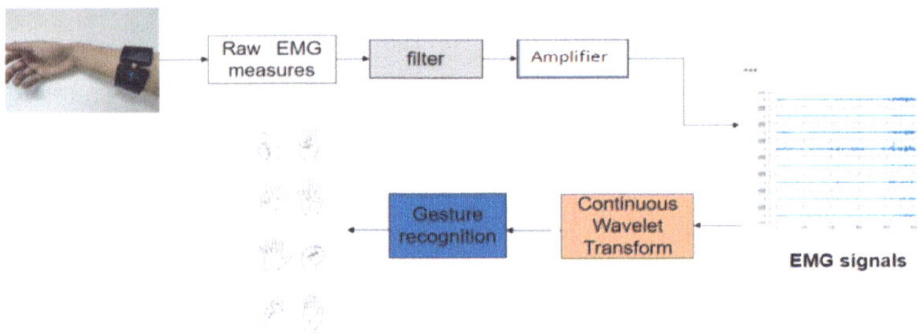

Fig. 3. Architecture of Gesture recognition system.

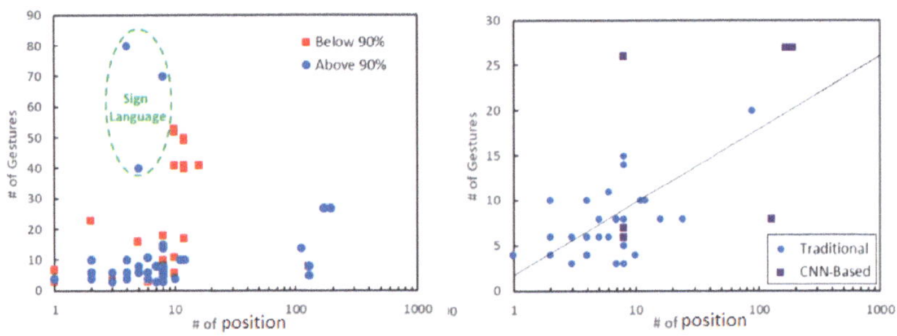

Fig. 4. Positions detected to the number of gestures given.

4 Conclusion

This article has effectively showcased the utility of EMG signals in the context of a human-machine neural interface. In contrast to similar research reliant on multiple EMG sensors, our proposed EMG-based control interface consistently demonstrates an impressive classification rate exceeding 97.5%, even with the use of just one EMG sensor. Furthermore, subjective assessments of the system's usability reinforce its robust performance. EMG signals offer valuable insights into the nervous system, and their adaptation for diverse applications is relatively straightforward. Additionally, the EMG data can capture static gestures or intentions through the identification of isometric muscle activation. Consequently, the EMG signal exhibits a broad spectrum of potential applications in human-machine interfaces. However, several challenges remain to be addressed in order to achieve advanced applications. These challenges encompass the development of specialized algorithms for specific EMG studies, the extraction of relevant features, and the creation of real-time classifiers that can be relied upon for precision, as outlined in this article.

References

1. Parveen, N., Roy, A., SaiSandesh, D., SaiSrinivasulu, J.Y.P.R., Srikanth, N.: Human computer interaction through hand gesture recognition technology. Int. J. Sci. Technol. Res. **9**(4), 505–513 (2020)
2. Veronica, P.G., Mokkapati, R.K., Jagupilla, L.P., Santhosh, C.: Static hand gesture recognition using novel convolutional neural network and support vector machine. Int. J. Online Biomed. Eng. **19**(09), 131–141 (2023). https://doi.org/10.3991/ijoe.v19i09.39927
3. Harshitha, J.L.V.S., Saranya, N., Sruthi, Y., Srithar, S., Chandra, J.: Hand sign classification techniques using supervised and unsupervised learning algorithms. In: 2023 International Conference on Computer Communication and Informatics, ICCCI 2023 (2023). https://doi.org/10.1109/iccci56745.2023.10128426
4. Pande, S., et al.: Assessment and recommendation of neural networks and precise techniques for sentiment systems analysis. J. Ambient Intell. Hum. Comput. **14**(8), 11285–11299 (2023). https://doi.org/10.1007/s12652-023-04643-4
5. Bhardwaj, S., et al.: Long-range technology-enabled smart communication: challenges and comparison. J. Circuits Syst. Comput. **32**, 2350161 (2022). https://doi.org/10.1142/s021812662350161x
6. Muzammil Parvez, M., et al.: Machine learning based models for defect detection in composites inspected by barker coded thermography: a qualitative analysis. Adv. Eng. Softw. **178**, 103425 (2023). https://doi.org/10.1016/j.advengsoft.2023.103425
7. Ahammad, S.K.H., et al.: Improved neural machine translation using natural language processing (NLP). Multimed. Tools Appl. **83**, 39335 (2023). https://doi.org/10.1007/s11042-023-17207-7
8. Jahan, K., et al.: Design and analysis of high speed multiply and accumulation unit for digital signal processing applications. Int. J. Recent Innov. Trends Comput. Commun. **11**(1), 95–102 (2023). https://doi.org/10.17762/ijritcc.v11i1.6055
9. Reddy, S.S., et al.: Hybrid autonomous vehicle (aerial and grounded). Int. J. Recent Innov. Trends Comput. Commun. **11**(1), 103–109 (2013). https://doi.org/10.17762/ijritcc.v11i1.6056
10. Sreelakshmi, D., Inthiyaz, S.: Fast and denoise feature extraction based ADMF–CNN with GBML framework for MRI brain image. Int. J. Speech Technol. **24**(2), 529–544 (2021). https://doi.org/10.1007/s10772-020-09793-w
11. Murali Krishna, B., et al.: FPGA implementation by using XBee transceiver. Indian J. Sci. Technol. **9**(17), 93032 (2016). https://doi.org/10.17485/ijst/2016/v9i17/93032
12. Suresh, N., et al.: Crop yield prediction using random forest algorithm. In: 2021 7th International Conference on Advanced Computing and Communication Systems, ICACCS 2021 (2021). https://doi.org/10.1109/icaccs51430.2021.9441871
13. Inthiyaz, S., et al.: Design of bi-trigger sram using schmitt trigger for low power 13t cmos application. Int. J. Sci. Technol. Res. **8**(12), 1466–1471 (2019)
14. Siva Kuma, M.: Delay estimation of different approximate adders using mentor graphics. Int. J. Adv. Trends Comput. Sci. Eng. **8**(6), 3584–3587 (2019). https://doi.org/10.30534/ijatcse/2019/141862019
15. Inthiyaz, S., et al.: Flower image segmentation: a comparison between watershed, marker-controlled watershed, and watershed edge wavelet fusion. ARPN J. Eng. Appl. Sci. **11**(15), 9382–9387 (2016)
16. Myla, S., et al.: Design decision taking system for student career selection for accurate academic system. Int. J. Recent Technol. Eng. **8**(9), 2199–2206 (2019)
17. Reddy, P.C., et al.: Customized digital logic circuit based design for physical unclonable function. J. Crit. Rev. **6**(5), 192–199 (2019)

18. Kumar, M.S., Inthiyaz, S., Babu, M.A., Teja, B., Ahmed, S.J., Harika, T.S.: Two speed radix-2 booth multiplier using verilog. J. Crit. Rev. **7**(04), 668 (2020). https://doi.org/10.31838/jcr.07.04.126
19. Jyothi, N.K., et al.: Synthesis and dielectric properties of PAN-based Na^+ ion polymer electrolytes. Mater. Today Proc. **51**, 1168–1171 (2022). https://doi.org/10.1016/j.matpr.2021.07.123
20. Inthiyaz, S., et al.: Facial expression recognition using KERAS. J. Phys. **1804**(1), 012202 (2021). https://doi.org/10.1088/1742-6596/1804/1/012202
21. Babu, K.S., et al.: Brain tumor segmentation of T1w MRI images based on integrated clustering algorithm. In: 2021 6th International Conference on Recent Trends on Electronics, Information, Communication and Technology, RTEICT 2021 (2021). https://doi.org/10.1109/rteict52294.2021.9573928
22. Kumar, M., et al.: Automatic origination of frequency division multiplexing in Xilinx-Vivado environment. In: Proceedings of the 10th International Conference on Signal Processing and Integrated Networks, SPIN 2023 (2023). https://doi.org/10.1109/spin57001.2023.10116550

Medrack: Bridging Trust and Technology for Safer Drug Supply Chain Using Ethereum and IoT

S. Porkodi(✉) and D. Kesavaraja

Dr. Sivanthi Aditanar College of Engineering, Tiruchendur, India
ishwaryaporkodi6296@gmail.com

Abstract. World of healthcare has been facing a revolution since pandemic and there is a huge rise in IoT devices developed for the betterment of healthcare sector. The drug supply chain plays a major role in enhancing healthcare, as volume of drugs and medical supply are increasing tremendously. Yet, complex drug supply chain faces various issues on counterfeiting and concerns are raised in terms of quality control, medicine hoarding and theft. The real time data from IoT sensors helps in tracking and monitoring drugs in containers during storage and during transportation throughout the supply chain from manufacturer till reaching the end customer. The immutability of the blockchain ledger ensures in providing the much-needed transparency and trust among the stakeholders. In this paper, a novel IoT application Medrack is built by utilizing blockchain as the back bone for the networked intelligent production system. The proposed work provides better solutions in solving the drug counterfeits, hoard problems and eradicates the theft also paving the way for deliver traceable and safe drugs to patients.

Keywords: Smart Contract · Drug Counterfeit · Supply Chain · IoT

1 Introduction

In the era of digitalization, both online and offline drug counterfeiting has become fastest growing illicit activity [1] affecting global population making drug supply chain crucial in healthcare industry. Statista [2] reports hike in pharmaceutical crimes up to 6000 global cases in 2021. In 2023, drug counterfeits of 20 million were recovered in Kolkata [3] and it is found that fake drugs are been manufactured in Himachal Pradesh and distributed throughout the country [4]. In more than 14 countries, there is rise in weight loss counterfeit drugs leading to dangerous health issues [5]. Even developed countries has counterfeit drugs in use risking the human lives. So, it is the end user's responsibility to verify the authenticity of the drug before consumption. The drug supply chain offers timely medicine delivery among end users where, the addition of blockchain adds authentication, ownership verification and transparency in tracking real time movement of drugs, creating trust while IoT adds real-time monitoring of factors such as tampering and temperature ensuring safe medications throughout the supply chain. All the data from IoT devices and transaction details of stakeholders via smart contract are recorded

V. E. Balas et al. (Eds.): IEQIP 2023, LNNS 1029, pp. 35–40, 2024.
https://doi.org/10.1007/978-3-031-61929-8_5

continuously in the blockchain with the timestamp in a chronological order (see Fig. 1). So, any stakeholder can be alerted in case of any theft or burglary or during emergencies when drugs are on demand and a particular stakeholder is hoarding it.

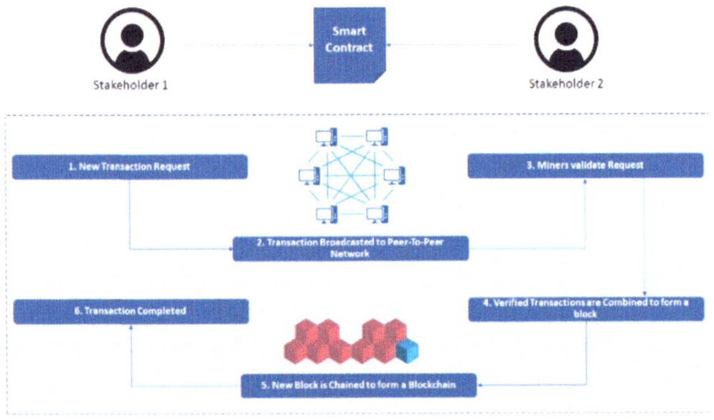

Fig. 1. Workflow of Blockchain

2 Related Works

In 2019, Bougdira et al. [6] proposed RFID tags and barcodes to identify and track goods. Environment data is captured with WSN in identifying product uniquely. However, the system does not have blockchain to store history of transactions permanently to maintain immutability. In the same year, Muniandy et al. [7] proposed a drug traceability system. Each drug is assigned ID & managed on the blockchain. If entered ID is not found, then drug is concluded as counterfeit product. The proposed work lacks detection of burglary and hoarder detection. Hewa et.al [8] proposed a telehealth system in 2020 connected with IoT and 5G based on permissioned blockchain and edge computing where the blockchain registers and monitors nodes. Smart contract is used to obtain permission from patient to access their medical data. In 2021, Musamih et al. [9] proposed blockchain based drug traceability system, where stakeholders are identified, blockchain is used to manage all transactions ensuring product safety and eliminating counterfeits with security and cost analysis is performed. But there are no IoT sensors or real-time location tracking to avoid burglary or vehicle hijack. Healthcare supply chain based on blockchain was proposed by Anusha et al. [10] in 2023 to trace drugs using blockchain in which Lot box containing drugs are identified using QR code. However, the proposed system does not support any IoT sensors to avoid burglary.

3 Proposed System

Stakeholders can be registered as nodes in Ethereum with smart contract after identity verification (see Fig. 2). Access control rights are given to stakeholders accordingly. Drug images at every stage are uploaded into decentralized storage along with unique

IPFS hash value. Data collected from stakeholders and their transactions with IPFS hash are stored in Ethereum. The stakeholders have access to both the Ethereum network and the decentralized storage to cross check the credibility of the product anytime.

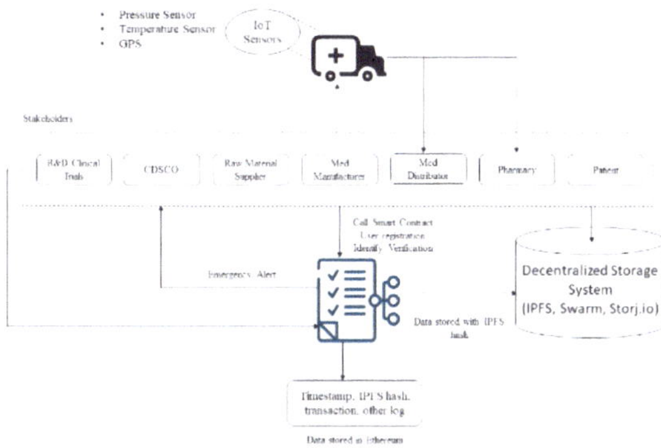

Fig. 2. Workflow of Medrack system

The IoT sensors present in medicine container of the delivery truck can provide GPS details, temperature and pressure values ensuring that medications are safe. If not, then the pre-set alert message is sent to the authorities with recorded parameters. Each recorded transaction with timestamp and the immutable data provenance is utilized to verify the credibility of drugs. If the details are not found then the drug is concluded as counterfeit drug. If a product is hoarded continuously by a person or a product remains in the warehouse for a very long time even in time of demand, then it can be detected and hoarder can be identified easily. The proposed system can be used to retrieve stolen truck carrying drugs with the GPS information. Buglers and thieves trying to steal medicine can be immediately reported if the vehicle deviates from the assigned paths. Transparency of the blockchain, timestamping of events and ownership verification in smart contract leverage trust among stakeholders consequently attaining trust.

4 System Implementation

Remix IDE is utilized to design smart contracts for user registration in which stakeholders are mapped to their unique wallet address as mentioned in Table 1 thus restricting any unauthorized access, each operation performed by the stakeholders (see Fig. 3) are carefully developed with modifiers preventing more than one operation happening at the same time consequently eradicating loopholes and reentrancy attacks then, access control policy specifies information on access rights for the stakeholders in the system.

The written smart contracts are compiled and tested by deploying in Goerli test network of Ethereum (see Fig. 4) where the log files can be analyzed to validate each transaction.

Oyente tool is utilized to scan for security vulnerabilities in the Medrack system before the smart contracts are deployed and the result displays false for 7 different vulnerabilities stating that the smart contracts are safe to use.

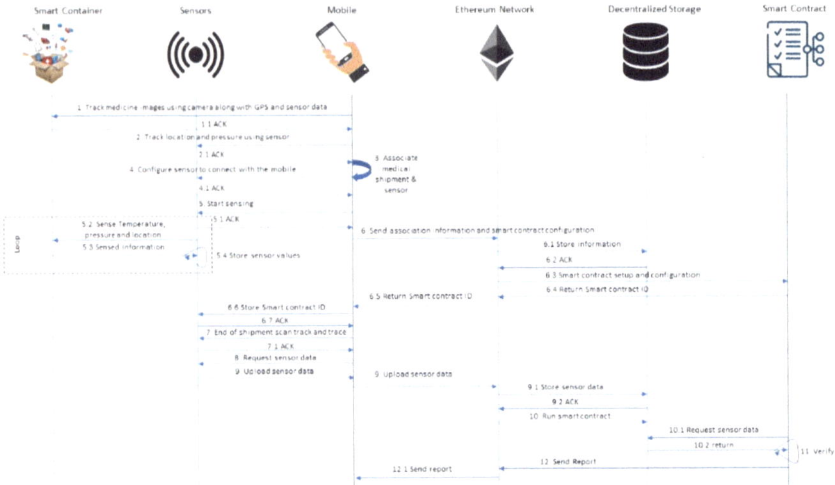

Fig. 3. Operations in Medrack System

Table 1. Stakeholders and their Wallet Address

Stakeholders	Wallet Address
R&D clinical trial (M τ)	0x5B38Da6a701c568545dCfcB03FcB875f56beddC4
M ϕ	0xAb8483F64d9C6d1EcF9b849Ae677dD3315835cb2
M γ	0x03C6FcED478cBbC9a4FAB34eF9f40767739D1Ff7
$M\mu$	0x4B20993Bc481177ec7E8f571ceCaE8A9e22C02db
M δ	0x78731D3Ca6b7E34aC0F824c42a7cC18A495cabaB
M ρ	0x617F2E2fD72FD9D5503197092aC168c91465E7f2
P	0xdD870fA1b7C4700F2BD7f44238821C26f7392148

Cost analysis is performed for the deployed contracts as each operation costs Ether as gas cost during contract deployment and transaction interaction as shown in Table 2. All the transaction cost are recorded in the real time on 15th Nov 2023 and can vary based on network traffic.

The total cost spent on contract deployment and interaction comes around a smaller Eth cost of 0.0139. Hence, trust management is achieved with the implementation of smart contract in Ethereum at such as smaller cost.

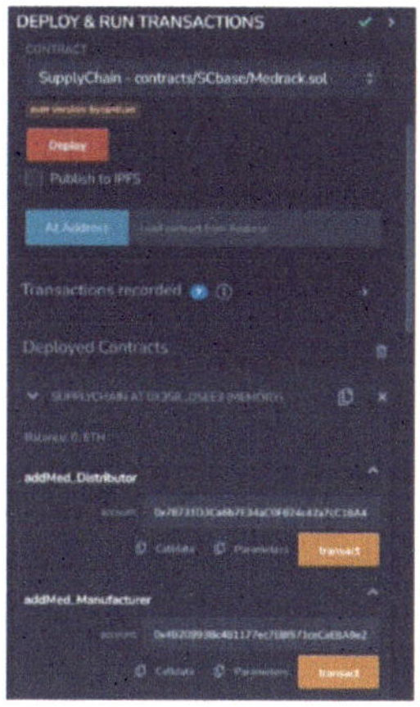

Fig. 4. Screenshot of the deployed Medrack supply chain

Table 2. Cost Analysis of Medrack System

SI. No.	Operation	Gas Cost	Real-Time Transaction Cost (Eth)
1.	Contract Deployment	3854536	0.00963634
2.	Stakeholders Registration	336917	0.00084231
3.	Operations	944537	0.00339145
Total		5135990	0.0138701

5 Conclusion

Thus, a novel application is built that shows advancement both in pharmaceutical and healthcare fields by utilizing the power of blockchain and IoT. Networked intelligent production system is developed for ensuring safety, authenticity, patient trust and improving drug supply chain. Smart contract with enhanced access control policy will auto stop unauthorized access, inside attacks and protects supply chain data. Blockchain makes system tamper resistant, trustable, auditable, traceable and secure withstanding various external cyber-attacks and reduce overall cost. The usage of IoT sensors prevents burglary and immediate actions can be taken on alerts from the intelligent system.

References

1. Ofori-Parku, S.S.: Fighting the global counterfeit medicines challenge: a consumer-facing communication strategy in the US is an imperative. J. Glob. Health **12**, 03018 (2022). https://doi.org/10.7189/jogh.12.03018x. PMCID: PMC9031510
2. Fleck, A.: Counterfeit Drugs on Rise Globally. Statista (2023). https://www.statista.com/chart/amp/30067/worldwide-counterfeit-pharmaceuticals-incidents/

3. Reuters, India: Drug Regulator Finds Counterfeit Medicines Worth 20 Million Rupees in Raid (2023). https://www.reuters.com/world/india/india-drug-regulator-finds-counterfeit-medicines-worth-20-mln-rupees-raid-2023-08-03/. Accessed 30 Oct 2023
4. Bhandari, R.: Major Bust in Himachal Pradesh's Pharma Hub Exposes Sophisticated Fake Drug Manufacturing Ring. The News Himachal (2023). https://thenewshimachal.com/2023/07/major-bust-in-himachal-pradeshs-pharma-hub-exposes-sophisticated-fake-drug-manufacturing-ring/
5. Wingrove, P., Fick, M.: Demand for Weight Loss Drugs Fuels Global Rise in Counterfeits (2023). https://www.reuters.com/business/healthcare-pharmaceuticals/demand-weight-loss-drugs-fuels-global-rise-counterfeits-2023-10-20/
6. Bougdira, A., Akharraz, I., Ahaitouf, A.: A traceability proposal for industry 4.0. J. Ambient Intell. Hum. Comput. **11**, 3355–3369 (2020). https://doi.org/10.1007/s12652-019-01532-7
7. Muniandy, M., Ong, G.T.E.: Implementation of pharmaceutical drug traceability using blockchain technology. Inti J. **2019**(35), 1 (2019)
8. Hewa, T., Braeken, A., Ylianttila M, and Liyanage, M.: Multi-Access edge computing and blockchain-based secure telehealth system connected with 5G and IoT. In: GLOBECOM 2020 - 2020 IEEE Global Communications Conference, Taipei, Taiwan, pp. 1–6 (2020). https://doi.org/10.1109/GLOBECOM42002.2020.9348125
9. Musamih, A., et al.: A blockchain-based approach for drug traceability in healthcare supply chain. IEEE Access **9**, 9728–9743 (2021). https://doi.org/10.1109/ACCESS.2021.3049920
10. Anusha, A., Shashidhar, T., Srivani, M., Srikanth, M., Ramesh, A.: A blockchain-based approach for drug traceability in healthcare supply chain. J. Eng. Sci. **14**(4), 1 (2023)

A Review on Tuberculosis Pattern Detection Based on Various Machine Learning Techniques

Devansh Agrahari, Akash Bhanushali(✉), Karunesh Biradar, and Chitra Bhole

Department of Computer Engineering, K.J. Somaiya Institute of Technology, Eastern Express Highway, Mumbai 400022, Maharashtra, India
{devansh.agrahari,akash.bhanushali,karunesh.b, chitra.bhole}@somaiya.edu

Abstract. This paper examines various respiratory tuberculosis detection methods, including conventional and enhanced algorithms, across multiple datasets. It evaluates the impact of image enhancement on the preliminary processing of TB chest Xray images and examines various image preprocessing strategies. Three image enhancing methods—Unsharp Masking, High-Frequency Emphasis Filtering, and Contrast Limited Adaptive Histogram Equalization— were evaluated to improve lung segmentation and heat mapping accuracy. The deep feature enhancement network identified in this study significantly improves accuracy, achieving 82.9% Average Precision and 77.6% recall on the TBXllK dataset. The paper also explores an advanced visualization technique that ensures comprehensive analysis and a deeper understanding of the subject matter. The CBAMWDnet model demonstrated outstanding performance in TB identification across multiple datasets, increasing the likelihood of early evaluation and treatment, marking a significant milestone in TB diagnosis.

Keywords: Tuberculosis · segmentation · image enhancement

1 Introduction

Chest radiographs are commonly used to identify tuberculosis in healthcare facilities, but they can be laborious and confidential, leading to incorrect diagnoses. The scarcity of competent radiologists and human limitations like fatigue and vision can also contribute to inaccurate diagnoses. Researchers have introduced "coordinate focus" to medical imaging, allowing convolutional neural network models to prioritize specific areas and extract features, making it flexible and effective in identifying and categorizing TB in chest X-ray pictures [1].

The recommended method uses pre-trained neural network models to detect pulmonary tuberculosis in chest X-ray images, enhancing efficiency and eliminating the need for manual feature extraction or lung segmentation [1]. Computer-aided diagnosis (CAD) systems are crucial for testing pulmonary tuberculosis through chest X-ray imaging. X-ray radiography offers low-cost technology and abundant data for deep learning methods. CNNs have been used to identify lung conditions from chest X-rays.

V. E. Balas et al. (Eds.): IEQIP 2023, LNNS 1029, pp. 41–46, 2024.
https://doi.org/10.1007/978-3-031-61929-8_6

CNN-based methods and Oxford's VGG framework were used to detect COVID-19 infections in 2020. CNNs are suitable for industries like monitoring systems, face recognition, automotive engineering, vehicle classification, and military [2] (Fig. 1).

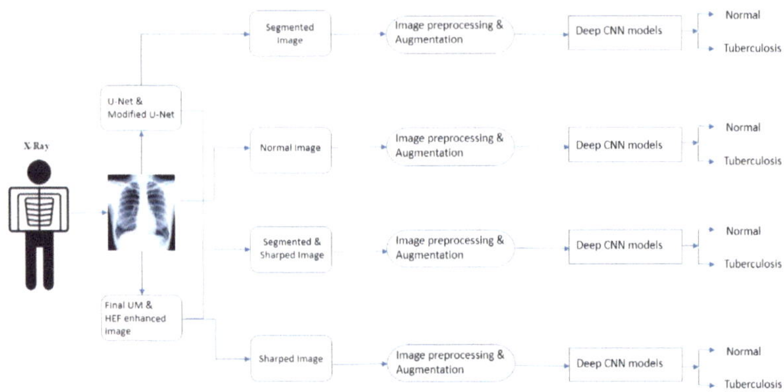

Fig. 1. Describes the proposed system design with a special emphasis on different processing techniques employed to improve the accuracy of the CNN model.

The main goal of this research is to assess the impact of two different preprocessing techniques on the performance of Convolutional Neural Networks (CNNs) trained to detect tuberculosis (TB) conditions [3]. On healthcare photographs, the Ultra Masking and HEF techniques have been demonstrated to be successful.

2 Related Work

2.1 Convolutional Based Learning and Support Vector Machine

Deep Convolutional Neural Networks (CNN) are widely used for image classification due to their ability to capture complex temporal and spatial features. Transfer learning reduces training time and maintains performance. Pre-trained CNN architectures like ResNet18, ResNet50, ResNet101, and MobileNetV2 are used for tuberculosis detection [3]. Support Vector Machine (SVM) is robust and precise for regression analysis and classification [4]. Kuang and Chen's method for evening vehicle detection uses nocturnal image enhancement, while Soomro's study explores retinal vein identification using deep learning.

2.2 Segmentation and Visualization

This study explores segmentation models using traditional U-Net architecture and its Modified U-Net, which uses rectification linear unit activations, 2×2 max pooling, and convolutional filtering techniques for more accurate results. It highlights the popularity of convolutional neural network models, such as Score-CAM, Grad-CAM++, and SmoothGrad, [5] in bridging the cognitive gap between machine learning and human cognition.

2.3 Feature Extraction

The study explores the impact of image enhancement on CNN models, revealing a significant increase in identification rates. Researchers used Local Binary Pattern (LBP) and Laplacian of Gaussian (LoG) [5] for image preprocessing, reducing noise, identifying edges, and tagging pixels for region-based analysis, particularly for lung areas like TBassociated abnormalities.

3 Proposed Solution

Tuberculosis, caused by Mycobacterium tuberculosis, is a common infectious disease causing mortality. Antibiotics are effective, and chest X-rays are used for diagnosis. A study using deep learning algorithms and segmented lungs can improve TB detection effectiveness. Nine deep neural network CNNs were trained to identify lung and non-TB diseases. Four experiments were conducted, including segmentation, image enhancement, and splitting X-ray images [6]. The goal is to develop transferable learningbased TB detection methods using active shape models.

4 Proposed Methodology

In this section, we will discuss the various methods that scientists use to accurately identify and distinguish TB patient's CXR (x-ray images) from a normal person. In this study we create texture and form features for a lung region to classify X-rays as

4.1 Data Acquisition

The TB database is derived from three publicly available datasets: Montgomery and Shenzhen datasets for drug resistance research, and the Belarus Set, containing 3,000 chest X-ray images from 3,087 tuberculosis patients.

4.2 Image Preprocessing and Enhancements

Image processing methods use dataset data to train convolutional neural networks (CNNs). X-ray images are resized and Z-score normalized to improve quality and safety. A study found that slight freezing improves categorization performance with VIT, VGG16, ResNet50, and VGG16-CoordAttention as backbone networks. VGG16 showed significant improvements, but MobileNetV2's smaller parameters and stop networks technique decrease classification performance.

4.3 Lung Segmentation

This study compares the performance of original and updated U-net channel segmentation models using metrics like Loss, Accuracy, IoU, and Dice. A new method for segmenting lungs in digital chest X-rays improves accuracy using patient-specific adaptive models.

Table 1. Comparative findings of various research papers describing the methods, dataset, and accuracy of methods achieved.

Author	Method	Database	Evaluation
Priyanka Saha	CNN- Dempster	3500 Itemsets	94.21%
Khairul Munadi	Deep Learning	Shenzhen Dataset	89.92%
Kahlil Muchtar	Deep Learning	Shenzhen Dataset	96.15%
Tawsifur Rahman	CNN Segmentation	7000 Itemsets	98.57%
Rahul Hooda	CNN	800 Itemsets	94.73%
An-Chao-Tsai	CNN	7950 Itemsets	AUC 0.996
Rosa Perez Siguas	Faster-RCNN	Not Defined	94.72%
Ramya R	Canny Edge Detection	Kenya Dataset	Not Defined
Shaoyue Wen	U-Net Neural Network	4000 Itemsets	91.2%
LLena Gabriella	CLAHE segmentation	Shenzhen Dataset	76%
Jaime Melen- dez	MiSVM technique	Zambia, Tanzania, Gambia Dataset	AUC 0.91
R.U.K Raja	Regression Analysis	Not Defined	AUC 0.953

Table 1 above provides an in-depth analysis of various tuberculosis detection methods, their datasets, and their corresponding accuracies, with each row representing a specific method normal or pathological, and evaluates its performance on two datasets, one from the county's health department's tuberculosis control program. [6].

4.4 Image Enhancement Algorithms

The study employs algorithms like sharp masking and high-frequency emphasis filtering to enhance the quality of TB CXR images before deep learning classification, assessing noise impact on sharpening features [7] (Fig. 2).

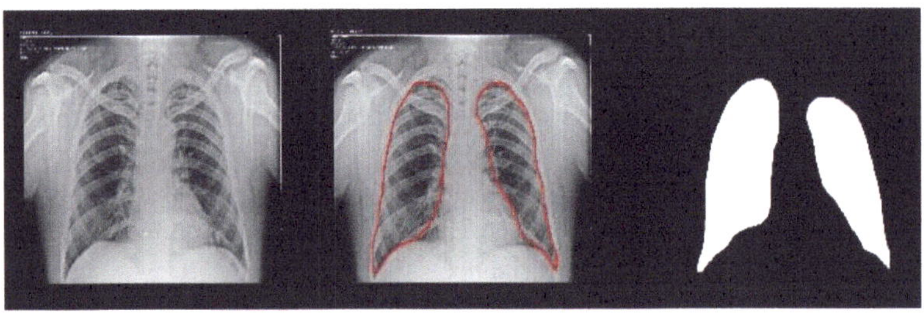

Fig. 2. Illustration of a classification dataset's X-ray image (on the left), mask generated through training the original U-Net model (in the middle), and associated medical condition (on the right).

This study uses newly introduced Score-CAM approach, a gradient-independent technique, to visually understand deep learning model decisions. It derives activation map weights by evaluating forward pass scores, determining the neural network's outcome by multiplying these weights with input image activation maps [8] (Fig. 3).

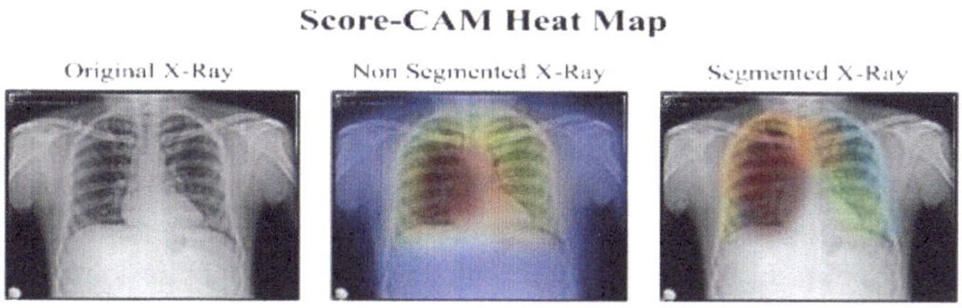

Fig. 3. Displaying Score-CAM heatmaps for correctly diagnosed tuberculosis-infected chest X-ray images: Original X-ray (left), Score-CAM heatmap overlaid on the original X-ray (middle), and ScoreCAM heatmap focused on segmented lung regions (right).

5 Result Discussion

The original U-Net architecture outperformed the modified U-Net in X-ray image segmentation, demonstrating strong segmentation performance. CheXNet, a transfer learning network, outperforms other convolutional neural networks in binary classification due to lower processing and training time [8].

6 Conclusion

A study has developed a transfer learning approach for automatically identifying tuberculosis in chest X-rays using convolutional neural networks (CNNs). The ChexNet model outperforms other deep CNN architectures in image segmentation and lung segmentation, demonstrating superior performance. The study focused on localized image analysis using the Laplacian form of Gaussian to describe lung tissue. The ChexNet model demonstrated high accuracy in tuberculosis detection, with segmentation enhancing classification accuracy. The study combined deep learning techniques with Unsharp Masking (UM) HEF image recognition to improve TB image recognition accuracy [9]. However, limitations like overfitting and computational resource requirements, privacy concerns, and bias towards specific demographics may affect results.

References

1. Santosh, K.C., Antani, S.: Automated chest x-ray screening: Can lung region symmetry help detect pulmonary abnormalities? IEEE Trans. Med. Imaging **37**(5), 1168–1177 (2018). https://doi.org/10.1109/TMI.2017.2775636

2. Bajaj, M., Rawat, P., Bhatt, A., Vats, S., Sharma, V.: A study on tuberculosis with deep learning and machine learning approaches. In: 2023 International Conference on Artificial Intelligence and Applications (ICAIA) Alliance Technology Conference (ATCON-1), pp. 1–6 (2023). https://doi.org/10.1109/ICAIA57370.2023.10169724
3. Capell´an-Mart´ın, D., G´omez-Valverde, J.J., Bermejo-Pel´aez, D., LedesmaCarbayo, M.J.: A lightweight, rapid and efficient deep convolutional network for chest x-ray tuberculosis detection. In: 2023 IEEE 20th International Symposium on Biomedical Imaging (ISBI), pp. 1–5 (2023). https://doi.org/10.1109/ISBI53787.2023.10230500
4. Gill, K.S., Sharma, A., Anand, V., Gupta, R., Deshmukh, P.: Influence of adam optimizer with sequential convolutional model for detection of tuberculosis. In: 2022 International Conference on Computational Modelling, Simulation and Optimization (ICCMSO), pp. 340–344 (2022). https://doi.org/10.1109/ICCMSO58359.2022.00072
5. Li, Y., Zhou, C., Wan, J., Wang, B.: Detection of tubercle bacilli by fusion with yolov5s and esrgan. In: 2023 IEEE 3rd International Conference on Information Technology, Big Data and Artificial Intelligence (ICIBA), vol. 3, pp. 659–663 (2023). https://doi.org/10.1109/ICIBA56860.2023.10165501
6. Ishtiaq, M.H., Rehman, F., Sarfraz, N., Sharif, H., Akram, H., Arshad, H., Manzoor, H.: Deep learning to predict pulmonary tuberculosis from chest poster anterior radiographs of lungs. In: 2022 14th International Conference on Mathematics, Actuarial Science, Computer Science and Statistics (MACS), pp. 1–5 (2022). https://doi.org/10.1109/MACS56771.2022.10022769
7. Huttunen, R., et al.: A comparison of signal combinations for deep learning-based simultaneous sleep staging and respiratory event detection. IEEE Trans. Biomed. Eng. **70**(5), 1704–1714 (2023). https://doi.org/10.1109/TBME.2022.3225268
8. Miranda, F.M., Fajardo, A.C., Medina, R.P.: Pulmonary tuberculosis detection using digitally photographed chest x-ray images. In: 2022 Second International Conference on Power, Control and Computing Technologies (ICPC2T), pp. 1–5 (2022). https://doi.org/10.1109/ICPC2T53885.2022.9776987
9. Hossain, S.I., Nipu, S.A., Hasan, M.R.: Recognition of tuberculosis on medical x-ray images utilizing mobilenet transfer learning. In: 2023 Third International Conference on Artificial Intelligence and Smart Energy (ICAIS), pp. 952–959 (2023). https://doi.org/10.1109/ICAIS56108.2023.10073728

Sensor Based Hand Gesture Identification for Human Machine Interface

V. Sri Sravan[1], Syed Inthiyaz[2](✉), and V. Rajesh[2]

[1] Department of CSE, Koneru Lakshmaiah Education Foundation, Vaddeswaram, Guntur 522302, A.P., India

[2] Department of ECE, Koneru Lakshmaiah Education Foundation, Vaddeswaram, Guntur 522302, A.P., India

syedinthiyaz@kluniversity.in

Abstract. This paper presents innovative and simple methods for extracting essential hand motion segmentation features from electrode signals obtained from the abductor longus muscle and human-machine interface (HMI) electromyography (EMG) signals in individuals with spinal cord injuries (SCI). The filtering process removed unwanted thumb-over-elbow sound, and distinct feature sets were derived. These feature sets are employed to control both the wheelchair's movements and an automated tool. Lastly, the study demonstrates that using SCI SEMG signals in an HMI, an individual can effectively control the cursor and navigate the wheelchair in the desired direction, resulting in a more user-friendly interface.

Keywords: Surface Electromyography (EMG) Signal Processing · Microcontroller · Microcontroller DAQ Card

1 Introduction

The necessity for more user-friendly interfaces and the humanization of machine-to-machine interactions is being stressed by professionals and academics in the disciplines of robotics and human-machine interface (HCI). Hardware and software solutions are required to accomplish this goal in order to protect users from the intricacies of technology in their workplace. It is essential to allow users to interact naturally with gestures and other inbuilt expressions. Because of this, the use of sign language and gestures into sophisticated interface design has been increasingly popular in recent times. The potential of gesture-based interfaces to improve and streamline a variety of tasks and applications through natural and intuitive handling is the reason for their increasing popularity. Human-Computer Interaction (HCI) technology uses bioelectrical signals, such as electromyography (EMG), electroencephalography (EEG), and electrooculography (EOG), to replace traditional input devices like keyboards and mice. The use of automated techniques based on EMG is crucial in the field of rehabilitation therapies. Bioelectric signals are used to create an intuitive and natural HCI. One important control source for this process is EMG, which records the electrical activity of muscles. Recording EMGs is a straightforward procedure as surface electrodes can be easily placed on

V. E. Balas et al. (Eds.): IEQIP 2023, LNNS 1029, pp. 47–53, 2024.
https://doi.org/10.1007/978-3-031-61929-8_7

the skin. The electrical signal generated by muscle contractions is known as EMG. Its applicability is confined to frequencies within the range of 0 to 500 Hz, and amplitudes that fluctuate between 0 to 10 mV (peak to peak). By utilizing differential gain, it is possible to effectively eliminate noise from the signal, allowing the EMG signal to be detected close to the skin's muscle. To achieve optimal results, it is advisable to position EMG electrodes along the muscle's longitudinal midline, typically between a motor point and the tendon insertion. These electrodes are commonly used on the extremities of patients dealing with conditions such as upper and lower spinal cord injuries and muscular dystrophy [1–4]. The primary goal of this innovative approach is to improve the level of service for individuals with disabilities, enabling precise control and navigation with minimal user involvement. In our research, we developed an automated program featuring a signal processing component capable of identifying motion commands based on hand gestures and utilizing signal conditions to trigger four distinct elements.

2 Methodology

2.1 Gesture Selection

The automated system relies on four distinct gestures for its operation. These gestures are designed to be easy to perform and should generate EMG signal patterns that are readily identifiable. Maintaining signal integrity involves positioning the hand in a specific manner referred to as the "starting position" both before and after each gesture. After evaluating more than 20 different gestures, we selected the four movements depicted in Fig. 1 as the control gestures because they best met the criteria outlined earlier, as shown in Table 1. The initial motion, labeled as "Gesture 1," involves a swift and wide hand motion. Following this, the second gesture, named "Hand Grab" (Gesture 2), is executed with a light and brief push. "Pinch Position" (Gesture 3) entails extending the fingers, while "Wrist Flexion" (Gesture 4) involves an inward turn of the wrist [5]. Our results reveal that the EMG waveforms generated by both hands are nearly identical, with minimal differences in overall amplitude. Consequently, distinguishing which hand is performing the action is not a challenging task. It's important to appropriately adjust these titles and descriptions for left-handed users [6].

2.2 System Structure

This section outlines the various processing stages of the system, as illustrated in Fig. 2, starting from gesture capture and extending to HCI option selection. The process initiates with electrodes attached to a sensor, which records the EMG signal originating from the working muscles of the arm. Subsequently, the signal undergoes preprocessing to prepare it for further analysis. The following step involves matching incoming patterns with corresponding gesture motions based on the signal. Initially, the distinctive features of each model are extracted to facilitate differentiation [7, 8].

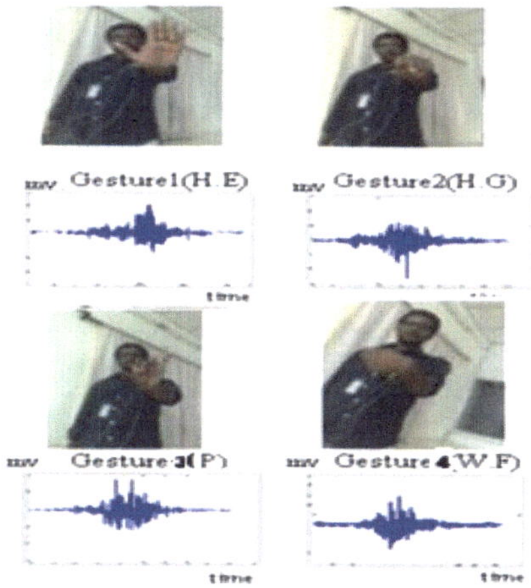

Fig. 1. Performance of gestures and the corresponding patterns.

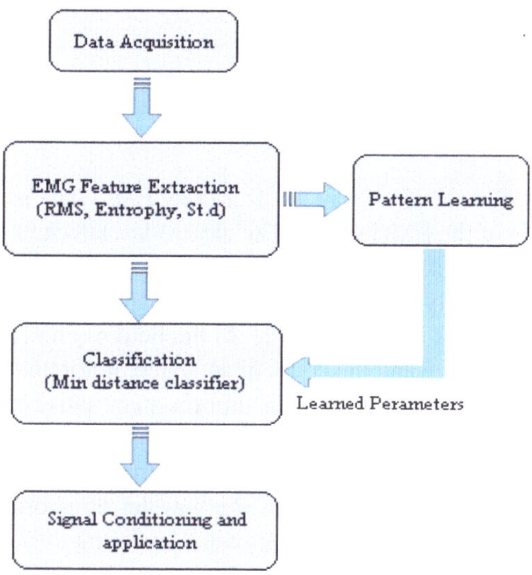

Fig. 2. A diagram outlining the analysis of EMG signals and pattern classification.

2.3 Signal Acquisition

In this particular setup, we utilized a surface EMG sensor to capture EMG signals. These signals exhibited amplitudes of up to 6000 μV and fell within a frequency range of 20 to 500 Hz. For signal acquisition, two single pre-gelled Ag/AgCl electrodes were applied to the skin along the abductor longus muscle of the system user, as shown in Fig. 3. To avoid unwanted interference or crosstalk from nearby muscles, each electrode pair was dedicated to the analysis of a single muscle. Due to hardware limitations, the setup allowed the analysis of only one sensor channel, enabling the simultaneous assessment of multiple muscles. The muscles primarily observed in this study were the flexor digitorum superficialis, responsible for finger movements, and the flexor carpi radialis and palmaris longus, controlling wrist movements [9]. These three electrodes were positioned midway down the forearm, aligned with the direction of the muscle fibers. Notably, the first electrode, situated near the wrist, provided the most accurate measurements between the tendon insertions and motor points in the forearm [10]. The reference electrode occupied a central position within this configuration. The system sampled the EMG signal at a rate of 125 samples per second.

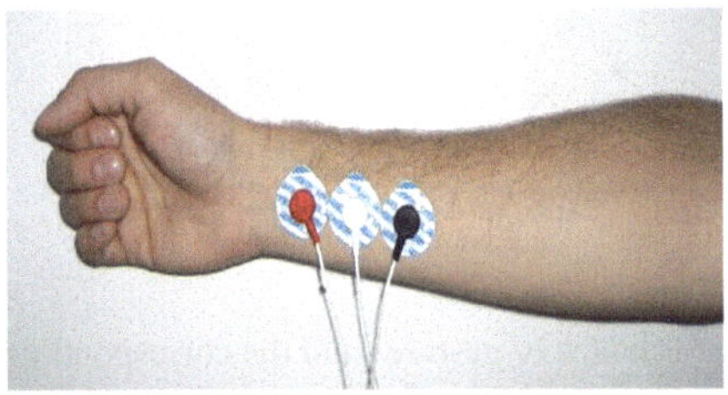

Fig. 3. Arrangement of electrodes based on the muscle structure.

2.4 Wavelet-Based Analysis of MUAPs

Decoding the operator's intentions directly from raw EMG data is a complex task due to the temporal nature of the EMG signal. The electrodes, attached to muscles, record a combination of artifacts, background noise, and action potentials from individual motor units (MUAPs). To effectively represent the essential shapes of surface MUAPs, a limited number of wavelet functions are needed [11]. In the field of clinical electromyography (EMG), the MUAP waveform and the count of activated motor units are critical components of the EMG signal [12]. Clinical EMG requires quantitative analysis, and computerized EMG devices have been employed to address this need. Various techniques have been applied to achieve quantitative analysis in both the frequency and time domains. Specifically, employing the db4 wavelet in the wavelet transform provides excellent temporal resolution at the highest frequency scales, covering only four signal samples. This unique property enables the db4 wavelet to efficiently capture the transient major peak of the MUAP signal, achieving temporal resolution not attainable with the STFT [13].

2.5 Feature Extraction

Precisely categorizing a executed gesture entails the identification and extraction of specific distinguishing characteristics from each associated model (Figs. 4 and 5).

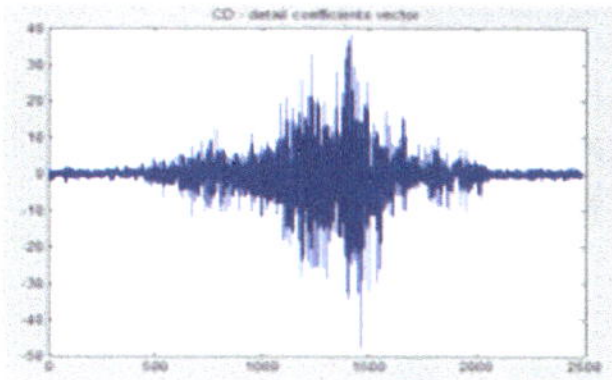

Fig. 4. Vector representing the coefficients.

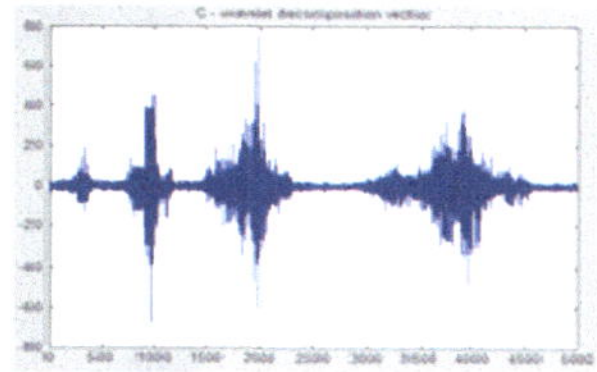

Fig. 5. Vector from wavelet decomposition.

2.6 Utilizing a Minimum Distance Classifier for Classification

The nearest neighbor technique aids in classifying unknown samples by assigning them to the same category as the most similar or "nearest" data point in the dataset. This method entirely circumvents the challenges associated with dealing with probability distances, as detailed in Instructions [14, 15]. In an n-dimensional feature space, the nearest distance corresponds to the minimum Euclidean distance. In our study, we utilized the minimal distance classifier based on the apple distance approach.

2.7 Experimental Setting

A total of twenty participants were involved in a long-term experiment. Initially, we collected biographical information from the respondents, including their age (with an average age of 21.85 years), gender (comprising nine females and eleven males), average weight (with an average weight of 52.15 kg), and dominant hand (nine left-handed individuals and twenty right-handed individuals). The experiment consisted of two stages. The first phase involved ten subjects, each performing six hand position gestures, resulting in a total of 60 recorded gestures. For each gesture, we calculated the mean of three feature values. These mean features were then used as test features for the initial ten participants. Subsequently, the remaining ten participants performed six additional hand gestures, which were set aside as test samples.

3 Results

Based on our earlier findings, we have identified entropy as the parameter that yields the highest detection rate for feature categorization. Through a visual analysis of the data, we have ascertained that entropy is the most effective feature for categorizing hand gestures. As depicted in Fig. 6, the entropy value is unique to the signal and serves as the target value for four distinct applications, including devices such as wheelchairs and mechanized tools (Fig. 7).

Fig. 6. HCI for SCI based on SEMG signal using Microcontroller.

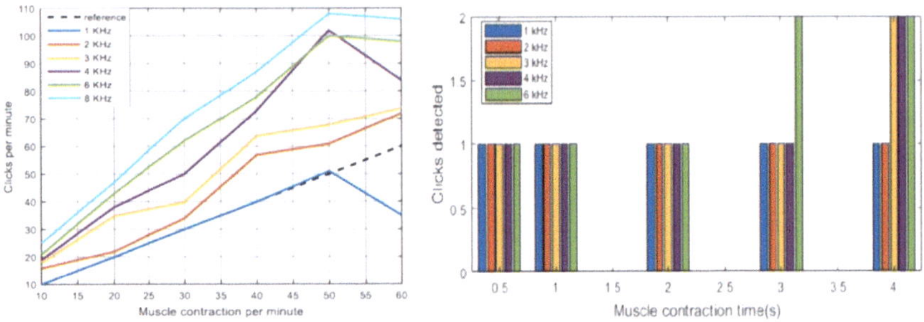

Fig. 7. Analysis of muscle contractions and Clicks per minute.

4 Conclusion

This study has effectively utilized an EMG signal within a neural interface, establishing a connection between humans and machines. The EMG signal, rich in information about the nervous system, proves adaptable for various peripheral applications with relative ease. Additionally, its capacity to detect isometric muscle activity enables the identification of intentions or motionless gestures within the EMG signal, presenting numerous potential applications in human-machine interfaces. Despite these achievements, there are several challenges that need addressing for advanced applications, including the development of real-time classifiers delivering consistent and accurate results, extraction of relevant features, and the design of algorithms tailored for specific EMG studies, as elaborated in this article. It is recommended to establish a standard practice of employing a compact, portable device that seamlessly integrates signal analysis with application functionalities in a single setup.

References

1. Rashed, A.N.Z., et al.: Spatial single mode laser source interaction with measured pulse based parabolic index multimode fiber. J. Opt. Commun. **0**(0) (2022). https://doi.org/10.1515/joc-2021-0214
2. Kumar, D., et al.: A test architecture design for SoCs using ATAM method. Int. J. Electr. Comput. Eng. **10**(1), 719 (2020). https://doi.org/10.11591/ijece.v10i1.pp719-727

3. Smirani, LK, et al.: Optical transceiver communication systems performance evaluation based on nonlinear cross gain modulation effects. J. Opt. Commun. **0**(0) (2022). https://doi.org/10.1515/joc-2022-0134
4. Sahoo, S.K., et al.: An optimized deep learning approach for the prediction of social distance among individuals in public places during pandemic. New Gener. Comput. **41**(1), 135–154 (2023). https://doi.org/10.1007/s00354-022-00202-1
5. Deb, P.P., et al.: A decision-making model with intuitionistic fuzzy information for selection of enterprise resource planning systems. IEEE Trans. Eng. Manag. **71**, 1–15 (2022). https://doi.org/10.1109/tem.2022.3215608
6. Smirani, L.K., Abd El-Aziz, I.A., et al.: Low loss flexibility and high efficiency of radio per fiber system for modern wireless communication system. J. Opt. Commun. **0**(0) (2022)
7. Ahammad, S.K.H., et al.: Defected ground-structured symmetric circular ring antenna for near-field scanning plasmonic applications. Plasmonics **18**(2), 541–549 (2023)
8. Zuhayer, A., et al.: A gold-plated twin core D-formed photonic crystal fiber (PCF) for ultrahigh sensitive applications based on surface plasmon resonance (SPR) approach. Plasmonics **17**(5), 2089–20101 (2022). https://doi.org/10.1007/s11468-022-01700-0
9. Uniyal, A., et al.: Surface plasmon resonance biosensor sensitivity improvement employing of 2D materials and $BaTiO_3$ with bimetallic layers of silver. J. Mater. Sci. Mater. Electron. **34**(6), 466 (2023). https://doi.org/10.1007/s10854-023-09821-w
10. Rashed, A.N.Z., Altahan, B.R., et al.: Grating index difference (GID) and order based distributed feedback laser effects on coarse wavelength division multiplexing optical fiber systems. J. Opt. Commun. **0**(0) (2022). https://doi.org/10.1515/joc-2022-0115
11. Rashed, A.N.Z., Babu, G.H., et al.: Mach–Zehnder electro-optic modulators chirping management in high-speed data communication lightwave systems. J. Opt. Commun. **0**(0) (2022). https://doi.org/10.1515/joc-2022-0238
12. Rashed, A.N.Z., Dasi, S., et al.: Performance signature of transceiver communication system based on the cascade uniform fiber Bragg grating devices. J. Opt. Commun. **45**, 165 (2022). https://doi.org/10.1515/joc-2022-0260
13. Rashed, A.N.Z., Jahan, K., et al.: Simulative study on integrated optical multimode waveguides with guided beams based on the system standardization of elements. J. Opt. Commun. **0**(0) (2022). https://doi.org/10.1515/joc-2022-0251
14. Rashed, A.N.Z., et al.: Single fiber grating profile multi scan parameters for the optimization of grating parameters by using optigrating simulation. J. Opt. Commun. **0**(0) (2022)
15. Nagageetha, M., et al.: "Performance analysis of feedback-based error control coding algorithm for video transmission on wireless multimedia networks. J. Adv. Res. Dyn. Control Syst. **9**, 626–660 (2017)

An Improved Detection System Using Genetic Algorithm and Decision Tree

Samudrala Harshadh[1], Mallu Vijay Kiran Reddy[1], and A. Suresh[2(✉)]

[1] Department of Networking and Communications, SRM Institute of Science and Technology, Chennai, Tamil Nadu 603203, India
{sh8335,mk8707}@srmist.edu.in

[2] Department of Networking and Communications, School of Computing, College of Engineering and Technology, SRM Institute of Science and Technology, SRM Nagar, Kattankulathur, Chengalpattu, Chennai, Tamil Nadu 603203, India
prisu6esh@yahoo.com, suresha2@srmist.edu.in

Abstract. These days, intrusion detection systems (IDSs) are gaining a lot of attention as an essential component of system defense. To safeguard the network, intrusion detection systems (IDSs) gather data on network traffic from various locations inside the computer system or network. It is quite tough and takes a lot of effort to discern between typical and intrusive network traffic operations. To determine the order of the network connection intrusion, an analyst needs examine all the extensive and varied data. It thus requires a method to represent the current network traffic that can identify network intrusion. In this paper, a unique approach to utilize data mining techniques and evolutionary algorithms for machine learning to identify intrusion characteristics for IDS was developed. Classification using a generative algorithm of decision trees is the method used to produce rules. These rules can identify the features of an intrusion and then be used as preventative measures in the genetic algorithm. To prevent intrusions in addition to identifying their presence.

Keywords: Genetic Algorithm · Denial of service · probe Attacks · Decision Tree · NSL-KDD Dataset

1 Introduction

As a result of the Internet's recent rapid expansion, computer systems have become increasingly vulnerable to security assaults. Computer system security is still extremely challenging, even with all the technological advancements in information assurance. Consequently, while the real software systems are operating, unauthorized breaches happen. The conveniences that modern technology has provided us are appreciated, but computer systems are becoming more vulnerable to internal and external security threats [1]. A truly safe system is practically unachievable, even with several defenses in place. Consequently, network intrusion detection, which tracks network traffic and spots unwanted network access, hostile computer system attacks, and unusual network

V. E. Balas et al. (Eds.): IEQIP 2023, LNNS 1029, pp. 54–59, 2024.
https://doi.org/10.1007/978-3-031-61929-8_8

behavior, is growing in importance. An intrusion occurs when someone tries to obtain access to a computer system or does anything that is prohibited by law [2]. External and internal intruders can be separated into two categories. The first group includes those who attack a system using different penetration techniques and do not have permission to access it [3]. It is the latter that describes those who are permitted to access yet want to do things that are not authorized. An accurate way to identify intrusions and notify the appropriate authorities about them is using an intrusion detection system [4]. Intrusion detection systems fall into two main categories: anomaly detection and abuse detection. While anomaly detection systems spot departures from typical network activity and warn of possible unknown threats, misuse detection systems find hackers using patterns they are familiar with. Some IDS create hybrid detection systems by combining anomaly and misuse detection [5]. Depending on where they search for intrusions, the IDSs may also be divided into two groups. Many methods for identifying network breaches have been developed, all based on soft computing. Soft computing is the term for a collection of methods that take use of tolerance for approximation, uncertainty, imprecision, and partial truth to achieve resilience and cheap solution costs. Soft computing approaches for intrusion detection are frequently combined with rule-based expert systems that express information as a series of if-then rules to acquire expert knowledge. Signature detection and anomaly detection are the two main categories into which intrusion detection techniques fall. If one examines the established intrusion patterns in system audit flow, every known kind of attack (intrusion) may be found using a misuse detection system. To detect anomalies, the system first establishes a profile of typical behavior before flagging any system events that deviate from this baseline. When it comes to identifying new or unexpected threats, the primary benefit of abuse detection is its high detection rate.

2 Literature Survey

The paper "UIDS: A Unified Intrusion Detection System for IoT Environment" by Kumar, V., Das, A. K., proposes a novel approach to detect and defend against various security threats in IoT networks. The authors begin by discussing the limitations of existing intrusion detection systems (IDS) and the need for a unified approach that can handle different types of attacks [6]. Neha Gupta, Vinita Jindal, and Punam Bedi have developed LIO-IDS, an innovative approach for addressing class imbalance in intrusion detection systems. These impressive results make LIO-IDS a prime candidate for real-world network deployment [7]. Nishtha Kesswani delves into the vital role of intrusion detection systems (IDS) in bolstering network security [8]. Hamizan Suhaimi introduces a cutting-edge strategy for identifying malicious connections within computer networks [9]. Witness the innovative work of Sydney Mambwe Kasongo. She has developed a cutting-edge intrusion detection system (IDS) tailored specifically for Industrial Internet of Things (IIoT) networks [10]. Syurahbil delves into the world of Intrusion Prevention Systems, specifically focusing on a technique called Decision Tree Data Mining [11]. Feng Gao introduces a groundbreaking approach to semi-supervised anomaly detection, highlighting the utilization of a limited number of positive samples [12]. Innovative and groundbreaking, YUANYI CHEN has revolutionized road anomaly detection with a unique approach that utilizes scale-invariant features [13].In their groundbreaking paper, "AnoViT: An Innovative Approach to Image Anomaly Detection," Yunseung Lee and Pilsung Kang present an unsupervised technique for detecting anomalies in images [14]. Hongju Cao expertly explores the utilization of Orthogonal Subspace Projection (OSP) in the dynamic field of hyperspectral anomaly detection. The ultimate findings of the research demonstrate that the OSP-based anomaly detection (OSP-AD) method surpasses its counterparts, solidifying its superiority in the field [15].

3 Methodology

The development of an enhanced intrusion detection system utilizing computational techniques will commence with the collection of network traffic data. This data will be obtained from diverse sources within the computer system or network, including network packets, logs, and other pertinent sources. This data will encompass both regular and potentially intrusive network traffic operations. Next, the collected data will undergo preprocessing, which involves cleaning and organizing it for analysis purposes. This process will include data cleansing, feature extraction, and transformation to ensure reliability and precision. At the core of our project is the application of machine learning techniques and evolutionary algorithms for intrusion detection. By incorporating classification methods like decision trees, we will generate rules that pinpoint intrusion features. Future work recommendations will also be provided, including the investigation of novel machine learning techniques and the incorporation of additional data sources for enhanced accuracy and reliability. In Fig. 1 shows the architecture diagram of the application. This data collection phase is akin to gathering information about what's happening in the computer system or network. First, we need to 'train' our system to understand what normal network behavior looks like using machine learning techniques, especially 'decision trees.' Decision trees are like roadmaps that guide us through the data, helping us create rules to spot unusual behavior. We fix inconsistencies, handle missing data, and ensure that everything is in a consistent and understandable format. Think of this as identifying the key details that help us distinguish between normal and abnormal network behavior during 'testing.'

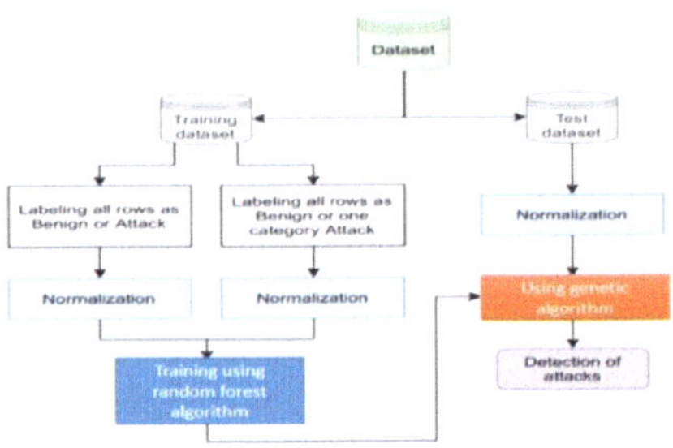

Fig. 1. Architecture diagram

These 'decision trees' play a vital role in understanding and classifying network behavior. And as an added layer of intelligence, we incorporate a 'genetic algorithm.'

It is as if our system is constantly 'learning' and 'adapting' to new challenges in the network. If the system 'spots' something suspicious during its 'testing' phase, it raises the alarm. We also create reports to summarize how well the system is working.

4 Discussion

In the case of intrusion, the decision tree and genetic algorithm are well trained to give optimal and efficient solution, the decision tree where the parent node acts as an intrusion and its child code acts as the decision by the parent weather it is an intrusion or normal entry. Firstly after installing the PyCharm we have aligned everything under the project interface, and by using python it makes us very easy to get the output which determines the intrusion. In the network. As we are working on anomaly-based intrusion with the help of dataset we have. we run the genetic decision tree and by that we get the precision, recall values by getting those values it will help it get the F1 score which will help to increase the efficiency and gives optimal and accurate result, and then we will run the other code which will help to divide the entries whether its normal or not.

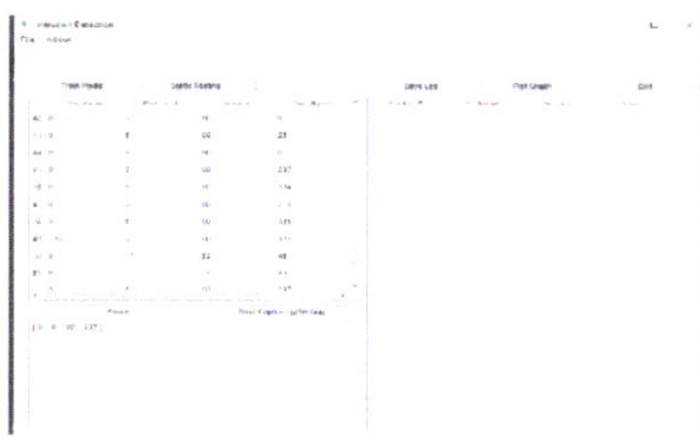

Fig. 2. Train and Test

In Fig. 2, we trained and tested our model on a comprehensive dataset. During the static testing phase, we closely examined both typical and malicious log entries from a popular dataset, known as nls-kdd on Kaggle. This robust dataset boasts almost 14 key factors, such as duration, protocol type, service, flag, source bytes, and destination bytes, among others.

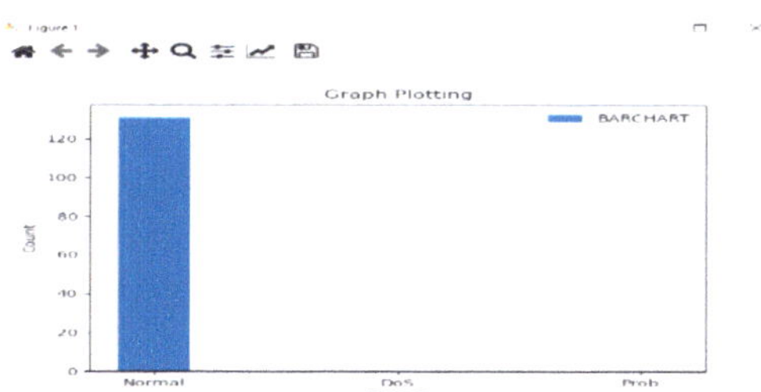

Fig. 3. Graph plotting

In Fig. 3 the graph plotting is done on the bases of the result obtained with the help of cross-checking the date base we have. Which helps us to determine whether the attack is normal are intrusion and again categorized into prob or dos attack. And reflect in the graph which can be saved as a file.

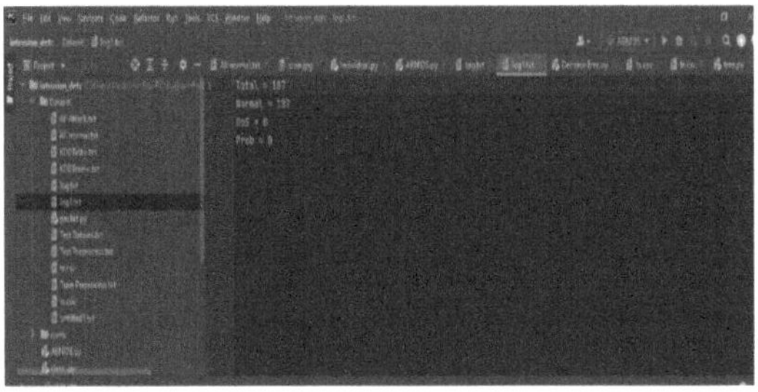

Fig. 4. Saved Log

In Fig. 4 illustrates the data which is been tested by the program the preprocessing and normalization of data and then the data is trained and tested with dataset that we have provided and integrated in the testing phase.

5 Conclusion

In this study, the focus is on intrusion detection, which is accomplished through the utilization of a genetic algorithm combined with decision tree methodology. The proposed approach is available in the form of software. Principal component analysis was employed to identify the critical features of network connections, while genetic algorithms were used to create classification rules for intrusion detection systems. The GA-based approach is characterized by its ease of implementation and maintenance, thanks to its straightforward rule representations and effective fitness functions. This system's flexibility allows it to be deployed across various application scenarios, provided that the right training dataset and attack classification system are accessible. Intrusion detection primarily aims to promptly identify and thwart attacks in real time, rendering threat classification irrelevant.

References

1. Yoon, S., et al.: Attack graph-based moving target defense in software-defined networks. IEEE Trans. Netw. Serv. Manag. **17**(3), 1653–1668 (2020)
2. Shen, H., et al.: Counterfeit anomaly using generative adversarial network for anomaly detection. IEEE Access **8**, 133051–133062 (2020)
3. Shajari, M., Geng, H., Kaixuan, H., Leon-Garcia, A.: Tensor-based online network anomaly detection and diagnosis. IEEE Access **10**, 85792–85817 (2022). https://doi.org/10.1109/ACCESS.2022.3197651

4. Kim, H., Lee, B.S., Shin, W.-Y., Lim, S.: Graph anomaly detection with graph neural networks: current status and challenges. IEEE Access **10**, 111820–111829 (2022). https://doi.org/10.1109/ACCESS.2022.3211306
5. Lundström, A., et al.: Improving deep learning based anomaly detection on multivariate time series through separated anomaly scoring. IEEE Access **10**, 108194–108204 (2022)
6. Kumar, V., Ayan, K.D., Diti, P.S.: UIDS: a unified intrusion detection system for IoT environment. Evolut. Intell. **14**, 47–59 (2021)
7. Gupta, N., Jindal, V., Bedi, P.: LIO-IDS: Handling class imbalance using LSTM and improved one-vs-one technique in intrusion detection system. Comput. Netw. **192**, 108076 (2021)
8. Ayyagari, M.R., et al.: Intrusion detection techniques in a network environment: a systematic review. Wirel. Netw. **27**, 1269–1285 (2021)
9. Suhaimi, H., et al.: Network intrusion detection system by using genetic algorithm. Ind. J. Elect. Eng. Comput. Sci. **16**(3), 1593 (2019)
10. Kasongo, S.M.: An advanced intrusion detection system for IIoT based on GA and tree based algorithms. IEEE Access **9**, 113199–113212 (2021)
11. Syurahbil, S., et al.: Intrusion preventing system using intrusion detection system decision tree data mining. Am. J. Eng. Appl. Sci. **2**(4), 721 (2009)
12. Gao, F., et al.: Connet: deep semi-supervised anomaly detection based on sparse positive samples. IEEE Access **9**, 67249–67258 (2021)
13. Chen, Y., et al.: Toward practical crowdsourcing-based road anomaly detection with scale-invariant feature. IEEE Access **7**, 67666–67678 (2019)
14. Lee, Y., Kang, P.: AnoViT: unsupervised anomaly detection and localization with vision transformer-based encoder-decoder. IEEE Access **10**, 46717–46724 (2022)
15. Chang, C.-I., Cao, H., Song, M.: Orthogonal subspace projection target detector for hyperspectral anomaly detection. IEEE J. Select. Top. Appl. Earth Observ. Remote Sens. **14**, 4915–4932 (2021)

A Detailed Analysis of Colorectal Polyp Segmentation with U-Network

Talluri Shiva Ramakrishna[1](✉), Gamini Rama Sai Asritha[1], Nagendla Lakshman[1], D. S. Guru[2], and Kolla Bhanu Prakash[1]

[1] Department of Computer Science and Engineering, Koneru Lakshmaiah Education Foundation, Guntur, Andhra Pradesh, India
2000032218cse@gmail.com, drkbp@kluniversity.in

[2] Department of Computer Science and Engineering, Mysore University, Mysore, India

Abstract. Colonoscopy is the most important technique for detecting colorectal cancer and its precursors. High miss rates continue to be a problem, leaving many abnormalities undiscovered. Systems for computer-aided diagnosis (CAD) have been developed using machine learning techniques to increase the accuracy of colonoscopies. CAD systems can aid in the discovery and characterization of lesions by pointing out areas in the colon that may have gone unnoticed during endoscopic examinations. To analyze colonoscopy pictures, these systems employ several approaches, including feature extraction, image processing, and classification. CAD systems can aid in the discovery and characterization of lesions by pointing out areas in the colon that may have gone unnoticed during endoscopic examinations. To analyze colonoscopy pictures, these systems employ several approaches, including feature extraction, image processing, and classification. Additionally, real-time assistance from CAD systems during colonoscopy examinations can give doctors a second opinion and lower the possibility of missing lesions. Additionally, by tracking and keeping an eye on lesions over time, these systems can aid in the early identification of potential malignant growth.

Keywords: Colonoscopy · Segmentation · Cancer · Neural network · U-network

1 Introduction

The greatest cause of cancer-related fatalities worldwide is colorectal cancer. Early detection and accurate diagnosis of colorectal cancer are crucial for improving the prognosis and survival rate of patients [4]. Colorectal cancer detection and diagnosis heavily rely on medical imaging. Image segmentation is a crucial stage in the study of medical images. Which involves identifying and separating regions of interest from the background. Due to their capacity to automatically learn complicated features, deep learning algorithms has received a lot of interest recently in the realm of medical picture segmentation (Fig. 1).

Reducing cancer risk, there is still a chance of missing adenomas, with an estimated miss-rate of 6–27% [12]. Studies have shown that small, flat, and sessile during colonoscopies, polyps are more likely to be missed. [2]. Polyps can also be missed if they are

V. E. Balas et al. (Eds.): IEQIP 2023, LNNS 1029, pp. 60–65, 2024.
https://doi.org/10.1007/978-3-031-61929-8_9

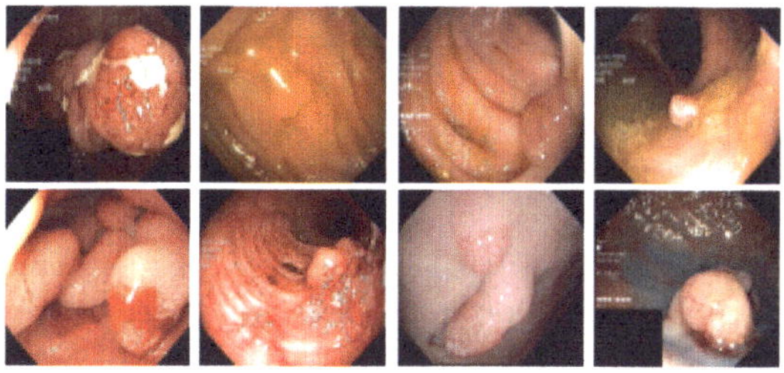

Fig. 1. An illustration of how polyps from the Kvasir-SEG can vary in size, shape, color, and appearance

not visible or not recognized due to the colonoscope being withdrawn too quickly [3]. To reduce the chance of missing adenomas during a colonoscopy, there are a few things that can be done. First, using the best observation methods and enhancing bowel preparation quality can both be beneficial. Additionally, it's crucial to allow at least six minutes of withdrawal time before a colonoscopy. [3].

2 Related Work

Researchers have been working on creating computer-assisted diagnostics prototypes for automated polyp segmentation for the past few decades. The edges or textures of the polyp were primarily examined in earlier approaches for segmenting polyps. Pre-trained networks and neural networks have been utilized in more recent techniques to improve segmentation accuracy. Bernal et al. [1] presented one such approach. They precisely identified polyps using WM-DOVA energy maps, regardless of their type or size. This allowed for more accurate localization of the polyp-Pozdeev et al. [8] also worked on developing a polyp segmentation approach using deep learning techniques. They used a CNN to segment the polyp and achieved promising results, demonstrating the effectiveness of deep learning in this field. Akbari et al. [11]. Used a specific type of artificial neural network called FCN-8S, along with a technique called Otsu's thresholding, to automatically identify and separate colon polyps in medical images. [13] Employed Seg-Net, a type of neural network, to find polyps in photos of the body. Specifically created for picture segmentation problems, Seg-Net [14] is a deep learning architecture. The researchers were able to achieve high levels of accuracy and performance, with a sensitivity and specificity that were both high, and a high value for ROC curve [9] Gastrointestinal medical pictures were analyzed for polyps using a Fully Convolutional Neural Network, a specific type of neural network. Their proposed method took first place in the high definition (HD) and standard definition (SD) image categories of the 2017 GIANA competition. In the 2018 GIANA challenge, the SD image segmentation job, their approach took second place. The FCNN model, which is a deep learning algorithm created especially for image mask problems, was successful in precisely recognizing and isolating polyps in medical pictures.

3 The U-Net Architecture

Architecture is characterized by its form architecture and was created for biomedical imagery. There are two paths: an enlarge path and a decline path, in the U-Network design. Convoluting and pooling layers make up the decline path, which reduces the resolution of the input image. Convol and up-sampling layers make up the expand path, which raises the output segmentation map's resolution. The U-Network architecture uses frisk connections between the corresponding layers of the declining and enlarge paths, which allow the network to preserve fine-grained details and localize features accurately. The network architecture is shown. The network has two parts: a declining path and an enlarge path. The declining path has repeated convolutions, ReLU activation, and pooling to down sample and add twice as many features channels. The enlarge path has up sampling, "up-convolution" It concatenates the corresponding feature map from the descending path with a reduction in the number of feature channels of half., and more convolutions with ReLU activation. Convolution is used in the last layer to map the feature vector to the required number of classes.

3.1 Encoder Block

The encoder block takes a contracting path, while the spatial resolution of the feature maps gradually decreases, the number of feature channels continually increases. This is accomplished by the sequential application of two 3×3 convolutional layers, each followed by Max pooling and a rectified linear unit activation function are then performed. The convolutional layers retrieve significant characteristics from the input image, whereas the ReLU activation function makes the network more nonlinear, allowing it to simulate complicated relationships between the input and output. The max pooling procedure decreases the feature maps' spatial resolution and down samples the input image.

3.2 Bottleneck

The bottleneck block, which joins the blocks for the encoder and decoder in the U-Network design, is a crucial part of the network. It is responsible for creating a brief representation of the image and retrieving key data from the input image that the decoder block will utilize to create the final segmentation map. Two 3x3 convolutional layers with a rectified linear unit activation function for each layer make up the bottleneck block in most cases There are the same number of channels across this block. While retaining the quantity of feature channels, the bottleneck block lowers the feature maps' spatial resolution.

3.3 Decoder Block

The decoder block in the U-Network design has the responsibility of up sampling the feature maps that the bottleneck and encoder blocks created to create the segmentation map's ultimate form. Building the final segmentation map is the decoder block's responsibility, and it is done by up sampling the feature maps that the encoder and bottleneck blocks created. And adding an up-convolutional layer that twice the feature map's spatial resolution. The up-convolutional layer is also known as the transposed convolutional layer or deconvolutional layer.

4 Experiments

4.1 Dataset

We have used CVC-Clinic DB dataset. The dataset does not divide into train, validate and test we later divided the dataset into train, validate, test in 80:10:10 formats. The information includes images of polyps as well as the real face behind the mask, which we use to predict the mask for polyps and assess the accuracy against the ground truth mask.

4.2 Evaluation Method

To evaluate the efficacy of polyp segmentation techniques, we use the metric Intersection over Union. These measures enable us to examine the similarity and overlap of the segmentation results obtained to the initial truth on the ground. While the amount of overlap between the two masks is determined by the IOU.

4.3 Implementation and Hardware Details

To implement all our models, The Keras framework and TensorFlow were utilized as the backend. To find the best set of hyperparameters for the U-Net model, we ran several tests on the same dataset with different sets of hyperparameters. Adam is the optimizer with an 8-batch size, binary cross-entropy is the loss function, and the learning rate is 1e-4, we discovered that the model performed effectively. The dice loss function proved competitive in our studies.

5 Results and Discussion

Due to its capacity to handle tiny datasets and deliver accurate segmentation results, The U-Network model has shown encouraging results in image segmentation, particularly in the study of medical images (Fig. 2).

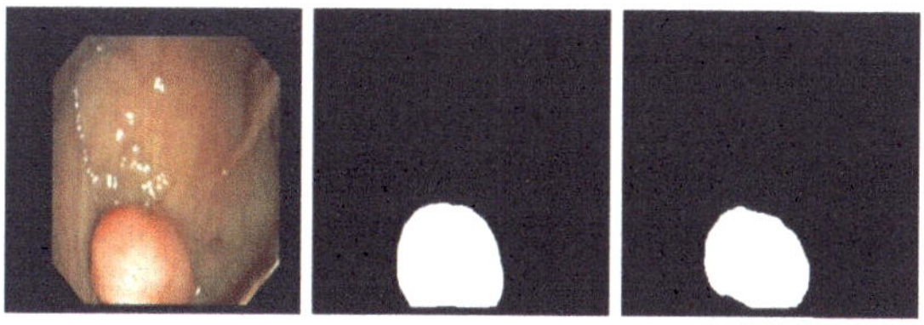

Fig. 2. Comparison between Ground Truth and model's predicted polyp shape

Table 1. Training Results

Variables	values
Validate Precision	0.8922
Validate Recall	0.6954
Intersectionover Union	0.4987
Validate Accuracy	0.9663
correctness value	0.9841
Recollect value	0.9011
Precision	0.9560
loss	0.0931
Validate IOU	0.3688

Table 2. Testing Results

Variables	values
Loss value	0.488
correctness value	0.9666
Recollect value	0.7887
Precision	0.8477
Intersection Over Union	0.4194

The U-Network model was also tested on multiple datasets utilizing various data augmentation approaches in order to improve the model's performance with the training sample. The model was built with the Keras framework and TensorFlow as a backend, and when given hyperparameters such an 8-batch size, it performed well, Binary cross-entropy is the loss function, Adam is the optimizer, and the learning rate is 1e4.

6 Limitation

1. Overfitting: U-Network is susceptible to overfitting, like many other deep learning models, especially when working with limited datasets. This can lead to poor generalization performance on unseen data.

7 Conclusion

Finally, specifically for medical image analysis, the U-Network model has proven to be a very efficient and widely utilized method for image segmentation tasks. Our results show that the U-Network architecture performs better than existing state-of-the-art algorithms for polyp segmentation tasks, with high Dice coefficients, Over Union, at the intersection and indicate intersection of values. From the union. The model's performance has been enhanced using data augmentation approaches, notably in resolving the issues of data scarcity and overfitting.

References

1. Pozdeev, A.A., Obukhova, N.A. and Motyko, A.A.: Automatic analysis of endoscopic images for polyps detection and segmentation. In: Proceeding of EIConRus, pp. 1216–1220 (2019)
2. Heresbach, D.O.: Miss rate for colorectal neoplastic polyps: a prospective multicenter study of back-to-back video colonoscopies. Endoscopy **40**(04), 284–290 (2008)

3. Vazquez, D., et al.: A benchmark for endoluminal scene segmentation of colonoscopy images. J. Healthcare Eng. **2017**, 1–9 (2017)
4. Bray, F., Ferlay, J., Soerjomataram, I., Siegel, R.L., Torre, L.A., Jemal, A.: Global cancer statistics 2018: globocan estimates of incidence and mortality worldwide for 36 cancers in 185 countries. CA Cancer J. Clin. **68**(6), 394–424 (2018)
5. Bernal, J., Sanchez, F.J., Fernandez-Esparrach, G., Gil, D., Rodrıguez, C., Vilarino, F.: Wm-dova maps for accurate polyp highlighting in colonoscopy: validation vs saliency maps from physicians. Comput. Med. Imag. Graph. **43**, 99–111 (2015)
6. Kang, J., Gwak, J.: Ensemble of instance segmentation models for polyp segmentation in colonoscopy images. IEEE Access **7**, 26440–26447 (2019)
7. Poomeshwaran, J., Santhosh, K.S., Ram, K., Joseph, J. and Sivaprakasam, M.: Polyp segmentation using generative adversarial network. In: Proceeding of EMBC, pp. 7201–7204 (2019)
8. Akbari,M. et al.: Polyp segmentation in colonoscopy images using fully convolutional network. In Proceeding of EMBC, pp. 69–72 (2018)
9. Yamada, M., et al.: Development of a real-time endoscopic image diagnosis support system using deep learning technology in colonoscopy. Sci. Rep. **9**(1), 1–9 (2019)
10. Ronneberger, O., Fischer, P., and Brox, T.: U-net: Convolutional networks for biomedical image segmentation. In: Proceding of MICCAI, pp. 234–241 (2015)
11. Wang, P., et al.: Development and validation of a deep-learning algorithm for the detection of polyps during colonoscopy. Nat. Biomed. Eng. **2**(10), 741–748 (2018)
12. Ahn, S.B., Han, D.S., Bae, J.H., Byun, T.J., Kim, J.P., and Eun, C.S.: The miss rate for colorectal adenoma determined by quality-adjusted, back-to-b
13. Badrinarayanan, V., Kendall, A., Cipolla, R.: Segnet: a deep convolutional encoderdecoder architecture for image segmentation. IEEE Trans. Patt. Anal. Mach. Intell. **39**(12), 2481–2495 (2017)
14. Guo, Y.B. and Matuszewski, B.: Giana polyp segmentation with fully convolutional dilation neural networks. In: Proceedings of VISIGRAPP, pp. 632–641 (2019)

A Review on Internet of Things (IoT): Parkinson's Disease Monitoring Device

J. Lokeshwari(✉) and R. Senthil Kumar

Department of CS with Cognitive Systems, Dr. N.G.P. Arts and Science College, Coimbatore, Tamil Nadu, India
lokeshwarimahanadhi@gmail.com

Abstract. The world is changing over time and is growing rapidly with various kinds of transformations while people are facing a lot of problems in their day-to-day life. The most common problem faced by people lies in healthcare sector. The Indian population statistics clearly states that the people of India lead a life up to the average age of 60 and then it is impossible to lead a life without any health issue. Not only from India, but also the people from various countries suffer with a lot of various ailments. One of the most common problems is tremor, due to tremor people are not able to carry out their daily routine. This happens because of nerve cell damage in the brain and it drops in dopamine levels, leading to symptoms of Parkinson's. Parkinson's is a type of disease that occurs due to the damage of the CNS and affects the action of a person or movement of a person and it leads to tremors. In this world millions of people are affected by this disease. When people are affected by Parkinson's disease, they have difficulties to do any kind of work with proper instructions because of tremors hence the individual needs an assistance from other persons to do their work. When the Parkinson's disease gets developed it reaches the final stage that makes an individual suffer badly. It is impossible to pay attention to watch Parkinson's patient throughout a day. When the patient reaches a final stage, one really needs a support and at the same time there will be a need to be motivated to recuperate from the disease. However, one has to monitor the patient through Internet of Things (IoT) assisted device. In this paper the devices for the tremors that are invented for Parkinson's disease is reviewed explicitly. In recent years IoT has made a technical revolution around the people by introducing user friendly devices it makes people highly comfortable to do their daily routine within a less time. It also makes the user to do the work faster and easier. The IoT is also advantaged in healthcare, industrial sector, small scale industries etc., Furthermore, this paper reviews on Parkinson's disease and IoT device usage in health care and also current developments in IoT assisted healthcare devices in real-time.

Keywords: Parkinson's Disease · IoT · WSN · IMU Sensor · Tremor · EMG device

V. E. Balas et al. (Eds.): IEQIP 2023, LNNS 1029, pp. 66–71, 2024.
https://doi.org/10.1007/978-3-031-61929-8_10

1 Introduction

An exploratory journey through the rapid pace of technological development in the health sector is significant. The cutting-edge developments like telemedicine, artificial intelligence, wearable medical equipment, and electronic medical records are swiftly increasing the way that healthcare is delivered. As the world learns to adapt to an era in which technology and healthcare are more closely linked than ever before. There is a need to explore the ethical, regulatory, and logistical challenges that are integrated with these technological advances. Through this investigation, we hope to get a thorough understanding of the seemingly limitless opportunities and potential problems that lie ahead in the rapidly developing field of technology-driven healthcare. Parkinson's disease is an extremely complicated and degenerative neurological disorder that has caught the attention of medical experts, doctors, and the general public. Parkinson's disease, with its distinctive symptoms of tremors, bradykinesia, stiffness, and postural instability, is not only an extreme challenge for people who are diagnosed. IoT devices are indisputably revolutionizing healthcare in the modern world by demonstrating real-time data, helping to improve patient outcomes, and boosting the effectiveness of healthcare. To determine that patients' best interests are protected as the technology develops.

2 Demand of Parkinson's Disease Monitoring Devices in World

From the year 1990–2019 according to the statistics of global trends Parkinson's disease is widely spread to 207 countries. In the national level the most three countries with highest percentage of this disease, are Qatar, the United Arab Emirates, Oman. This disease affects the mid –aged people and above 60 according to the latest data over 10 million people of the worldwide are affected by this disease. Due to this Parkinson's disease more than 5,847 deaths are occurring per year. When compared to Women, Men are mostly affected by this disease. Probably 10 men are affected compared to 3 women per month (Fig. 1).

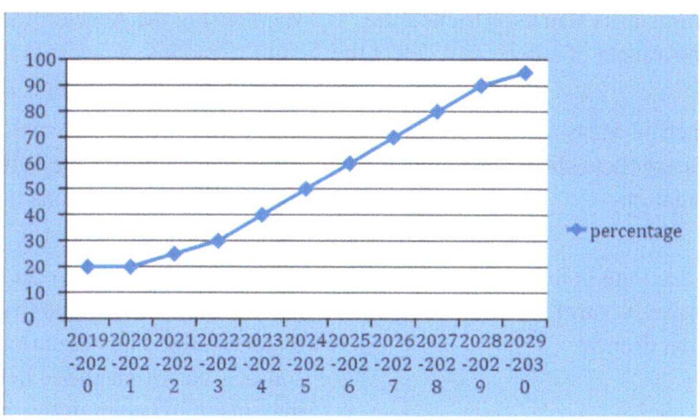

Fig. 1. Numbers of Peoples Affected by Parkinson's Disease

3 IMU (Inertial Measurement Unit) Sensor

IMU (Inertial Measurement Unit) is made with the combination of accelerometer and gyroscope. Accelerometer and gyroscope are sensors which are used to calculate the degree of a person's position and the movement of a person [1]. After calculating the Accelerometer and Gyroscope it identifies the affected person using IMU which is the position tracking system. When the IMU sensor is fixed on to a person and then it starts to measure the movement in frequency which is also used to measure a rotational position of a body in the angles of 130, 60, 90 etc. [2].

3.1 Directions of IMU (Inertial Measurement Unit) Sensor

IMU sensor is fixed with integrated Accelerometer and Gyroscope. The Accelerometer calculates the vibration and movements of a person. One can also measure the vibration on a micro-scale [3]. The main work of this sensor is to convert the vibration into a Piezoelectric effect. Piezoelectric is commonly used for the generation of charge and it works with stress and pressure [4]. Stress is also divided into two ways that are compressive stress and Tensile Stress. When stress is applied to conducting materials like silver, copper, it moves forward to the conducting surface. The General principle of Gyroscope is based on gravity and it is also a result of angular momentum [5]. This Gyroscope can move freely more than single directions. Gyroscope principle is also based on Newton's first law. This method is also performed in an internal hallway of the hospital. The experiment is carried out by making the participants walk for 15 m along the straight path and the same task is repeated thrice to check the Anthropometric measures of foot and the length of trunk [15] (Table 1).

Table 1. Advantages and Disadvantages of IMU sensor.

S. NO	Advantages	Disadvantages
1	Accelerometer is also used to measure linear movement of a person to detect the tremors	While using the Accelerometer sensor one cannot use this to measure the angular velocity but in Gyroscope. [6]
2	IMU (Inertial Measurement unit) also provides excellent short-term motion data or information	The device always measures the changes in contrast to itself. The IMU constantly rounds off small fractions in its calculations [1]
3	IMU is accurate in finding the result of Pitch angles, control tasks and also navigation drones	Sensors that are used to detect the disease harm one physically due to excessive temperatures. The disruption from various sources, electromagnetic fields, sensors may result in uncertainties. [8]

4 Keyboard for Hand Tremors

Steady type is a keyboard designed for people with hand tremors. Steady type is used to type relatively faster and accurate result [9]. Steady type can be also used for both IOS and Android because it is a third –party applications. Steady type is an on-screen Keyword. One should give the finger a little flexibility where it primarily hits the display and then focuses on what happens when the finger interacts with the screen.

4.1 Directions of Keyboard

The user ought to constantly receive the choice to abandon an activity without missing a mistake that needs to be remedied with an alternative action. This inhibits the agonizing process of blunders leading to additional mistakes. Similarly, constant touches and unintentional touches should not undermine the experience. The hit targets should be as massive and separated as one can. Accuracy can be compensated for some speed. This will be much faster as well as more enjoyable at the end. The smallest quantity of keystrokes possible should be used to type words and sentences [10]. Every keystroke demand works; one should make each one as efficient and feasible. One needs to be cautious while implementing timing dependent functionality. Tremors and other movement related issues may require people to move at their own pace. We should provide configuration options to the user. Customizability boosts user friendliness (Table 2).

Table 2. Advantages and Disadvantages of Keyboard.

S. NO	Advantages	Disadvantages
1	The steady type is also included as voice input	In iPhone OS it is difficult for the person with hand tremor to type by hand because IOS places have many restrictions on third-party keywords
2	Many tremor sufferers realize that they have better controls when touching something with hand even when their hand and fingers are at rest [11]	The main disadvantages are it is very tough to enter the passwords and telephone keypads Correcting an error necessitates an extra collaboration that may be more vulnerable to error
3	Stress always contributes to tremors, eventually leading to an apprehensive feedback cycle when typing something with a hand tremor or quiver	When we use the on-screen keyboards, it is very difficult even for normal people. However, the key is very small and it is placed closed together

5 EMG (Electromyography)

EMG (Electromyography) is often used to detect the muscle disorders that is polymyositis or muscular dystrophy [12]. EMG (Electromyography) has a low-risk procedure [1]. EMG (Electromyography) is a needle made of electrode that is ingested into a muscle of a human body directly and it starts to record the electrical activity in that muscle [13].This works, based on the combination of Accelerometer and surface EMG electrodes. In order to determine the neurological drive leading to pathological tremor oscillations, high-density EMG signals from the wrist flexors and extensors are acquired and processed [1]. Predicted stimulation timing for reducing tremor is based on the phase of difference between the neural drives and muscles. From 0 to 6 skins temperature is measured and it gets decreased when ingested into the muscle [14].

5.1 Directions of EMG (Electromyography)

Oxford University postulated a Medelec Synergy EMG equipment, it was put to use with four EMG surface electrodes and one pressure accelerator [13]. The muscles of the flexor and extensor capitularies were given a surface electrode application. In the same hand, the electrode was placed to the distal side of the third metacarpal bone. The frequency and amplitude of the patients' tremor were measured while they were resting with their hands on their thighs, and in a different posture with their arms crossed in front of them, and with their arms loaded with 500 g and 1000 g, respectively. The EMG signals muscle activity indirectly calculates not only physiological signals but also calculates the noise and artifacts [16] (Table 3).

Table 3. Advantages and disadvantages of EMG sensor

S. NO	Advantages	Disadvantages
1	EMG (Electromyography) is very useful to collect the information of the muscle's activity which is recorded by using needle, wire electrodes or more commonly surface electrodes	EMG (Electromyography)has most common disadvantage is that needle electrodes is directly inserted through skin into the muscle tissue of a human body, which causes pain while inserting and removing
2	EMG (Electromyography)is used to detect the electrical potential of the muscles cells when these cells are activated either electrically or neurologically [1]	EMG (Electromyography) commonly monitors nerve roots
3	It is also typically used to diagnose the neurogenic or myogenic diseases	Result will be less accurate compared to other tremors devices because of the environmental disturbances

6 Conclusion

This paper reviews the Parkinson's disease monitoring device and its distinctive symptoms of tremors, bradykinesia, stiffness, and postural instability, is not only a challenge for people who are diagnosed but also a field of study that illustrates the dynamic junctions of medicine, neuroscience, and technology. IoT devices are indisputably revolutionizing healthcare technology in the modern world by demonstrating real-time data, helping to improve patient outcomes, and boosting the effectiveness of healthcare. To emphasize that patients' best interests are protected as this technology develops.

References

1. Brindha, A., Sunitha, K.A., Robert Wilson, S.: Tremor classification using wearable IoT based sensors. IOP Conf. Ser. Mater. Sci. Eng. **1219**(1), 012024 (2022). https://doi.org/10.1088/1757-899X/1219/1/012024
2. Vescio, B., Quattrone, A., Nisticò, R., Crasà, M.Q.: Aldo, wearable devices for assessment of tremor. Front. Neurol. **12**, 680011 (2021)
3. Dai, H., Zhang, P., Lueth, T.C.: Quantitative assessment of parkinsonian tremor based on an inertial measurement unit. Sensors **15**, 25055–25071 (2015)
4. https://www.elprocus.com/
5. https://www.ceva-dsp.com/
6. Zhu, H., et al.: IMU motion capture method with adaptive tremor attenuation in teleoperation robot system. Sensors **22**, 3353 (2022). https://doi.org/10.3390/s22093353
7. Dayle, R., Sujitha, M., et al.: Tremor analysis with wearable sensors correlates with outcome after thalamic deep brain stimulation. Clin. Parkinson. Related Dis. **3**, 100066 (2020)
8. https://www.steadytype.com/
9. https://www.washingtonpost.com/
10. https://calahealth.com/
11. https://www.mayoclinic.org/
12. Zhang, J., Xing, Y., Ma, X., Feng, L.: Differential diagnosis of Parkinson disease, essential tremor, and enhanced physiological tremor with the tremor analysis of EMG. Parkinsons (2017). https://doi.org/10.1155/2017/1597907
13. Hess, C.W., Pullman, S.L.: Tremor: clinical phenomenology and assessment techniques. Tremor Hyperk. Movem. **2**, 02 (2012). https://doi.org/10.5334/tohm.115
14. Anbalagan, B., Karnam Anantha, S., et al.: A non-invasive IR sensor technique to differentiate Parkinson's disease from other neurological disorders using autonomic dysfunction as diagnostic criterion. Sensors (Basel). **22**(1), 266 (2021)
15. Caramia, C., et al.: IMU-based classification of Parkinson's disease from gait: a sensitivity analysis on sensor location and feature selection. IEEE J. Biomed. Health Inf. **22**(6), 1765–1774 (2018). https://doi.org/10.1109/JBHI.2018.2865218
16. Islam, A., Alcock, L., Nazarpour, K., et al.: Effect of Parkinson's disease and two therapeutic interventions on muscle activity during walking: a systematic review. Dis. **6**, 22 (2020)

Machine Learning-Based Prediction of Temperature Rise in Squirrel Cage Induction Motor (SCIM)

B. Sudha, Anusha Vadde(✉), and L. S. Praveen

Ramaiah University of Applied Sciences, Bangalore, India
vaddeanusha203@gmail.com

Abstract. The three phase SCIM plays an important role in automobile and production industries due to its efficient and reliable operation. These are workhorse of all industries. They are low cost motors. Heavy continuous duty, load conditions, different duty cycle, wrong installation, environmental stress, temperature dissipation and manufacture imperfections can affect the life of machine. The temperature dissipation in SCIM is the main trouble for malfunction in it. Now a days, temperature prediction on induction motor is become more important in machine design companies. There are many conventional methods to predict the temperature rise in induction motor, but it requires more time and more knowledge on simulation software. In the current work the temperature rise on three phase SCIM is predicted using Machine Learning (ML) algorithms. The results of simulation is recorded, this temperature rise data is used in ML. The best ML algorithm for predicting temperature rise in SCIM is K-NN algorithm. In the present research work the cause of temperature rise is studied and the correct ML classification algorithm is initiated for temperature prediction in SCIM.

Keywords: Squirrel Cage Induction Motor (SCIM) · Machine Learning (ML) · Temperature Rise

1 Introduction

The temperature rise in induction motor needs to be recorded at a proper time for reducing the loss [1]. ML algorithm applications in the prediction of temperature rise gives effective solution for healthful maintenance. To reduce losses in induction motors, it must be repaired at particular time [2]. ML algorithm gives an effective deterrent care solution. SCIM have increased popularity in the past decade. An examination of the increase in temperature within electric motors is crucial for ensuring their optimal operation and identifying any potential malfunctions [3]. Significant temperature increase can result in insulation degradation to the stator winding, perhaps leading to a short circuit. The enhanced temperature in the motor leads to the deterioration of bearings and the degradation of permanent magnets, resulting in a reduced lifespan of the machine. A temperature increase of 10 °C in a motor results in a 50% reduction in its lifespan [4–6]. SCIM identifies three fundamental kinds of thermal losses. The initial consideration is

V. E. Balas et al. (Eds.): IEQIP 2023, LNNS 1029, pp. 72–78, 2024.
https://doi.org/10.1007/978-3-031-61929-8_11

the copper winding losses, the magnitude of which is contingent upon the current flow. Additionally, there are core iron losses that are predominantly contingent upon the stator voltage. Mechanical losses constitute the final category and are factored into the motor speed. Insulation on stator windings is especially susceptible to temperature fluctuations and the thermal deterioration process [7]. It is challenging to measure temperature without implementing measuring sensors, which increases both the cost and weight of the motor. Electric drives have gained widespread adoption in the automotive industry, and ongoing research aims to decrease the motors' weight. In light of advancements in deep learning, various ML approaches have been suggested for the purpose of forecasting sensor values [8]. Insulation in stator windings is especially susceptible to the detrimental impacts of elevated temperatures and the thermal aging phenomenon. If the temperature rise of SCIM is not controlled, the heat will transferred to other components through casing and airgap. This will lead to bearings and permanent magnets heat. The main aim of researcher is to predict the temperature rise in SCIM, so all these losses can be reduced. The best and fast, modern and effective method for predicting temperature rise in SCIM includes ML and deep learning. The advantages of these methods is that, it will predict the temperature rise of SCIM on the basis of collected measurement data [9]. Several ML technics are existing and followed to predict the temperature rise in SCIM, like random forest, artificial bee colony, linear model, artificial neural network, decision tree, support vector method, Naïve Bayes, etc. These are used to predict the temperature rise in SCIM [10, 11].

Objectives of research;

- To collect temperature rise data of different parts of SCIM
- To classify temperature rise data using ML algorithm
- Producing confusion matrix for simulated data

The induction motor can be protected from high temperature distribution, by reducing the main losses like stator copper loss in stator winding, iron loss is rotor bars, eddy current loss in stator and by providing sufficient cooling system [12]. Before the conventional method is used to predict the temperature rise in SCIM. But conventional method needs more time to predict the temperature rise. In ML method, the temperature prediction is easy and fast [13]. The methodology of the work is depicted in block diagram as shown in Fig. 1. It shows that for process training data and testing data play an important role. Initially a 15 kW SCIM is selected and simulation on 15 kW SCIM is performed using finite element methods. From the simulation, temperature rise in different parts of 15 kW SCIM is recorded. The temperature rise data is recorded as training data and current, voltage, power, speed, frequency, torque, power factor, stator resistance, slip, loss and efficiency are taken as testing data. Lastly the confusion matrix is proved for best ML algorithm.

The mentioned ML algorithm is developed in MATLAB. The comparison of feature classifications and feature extractions results is given Table 3. The classified data is utilized by ML algorithms, SVM, decision tree, random forest, and linear model. Actually the temperature rise in different parts of SCIM is occurs due to increase in losses, there will be small change in parameters. In this paper the prediction of temperature rise in different parts of SCIM with speed, frequency, torque, power factor, stator resistance, slip, loss resistance variation due to various losses. Temperature rise data sets that have been recorded contain the following: frequency in Hertz, power in Watts, current in Amperes, voltage in Volts, and resistance in Ohms [14]. Based on the present values pertaining to speed and slip for various loads, this fault data is classified into four distinct classes: no load, half load, three quarter load, and full load. The first class is no-load, the second is half load, the third is three-fourth load, and the fourth is full load.

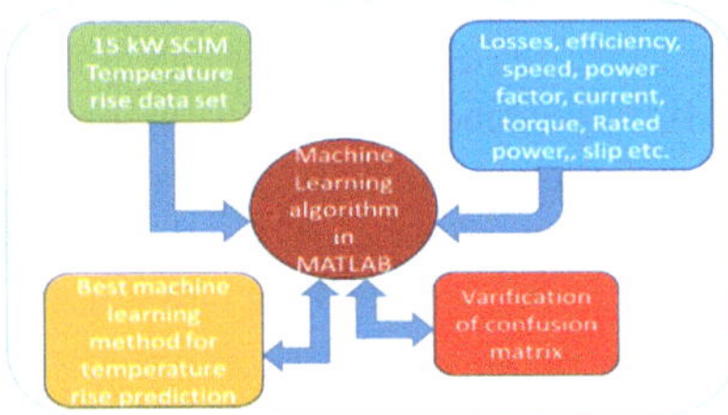

Fig. 1. Process block diagram

2 Simulation Data Set of Temperature Rise in SCIM

The materials used for design of induction motor are subjected to deterioration from thermal stresses, electrical stresses and mechanical stresses. The insulation breakdown is presence in the motor due to the thermal excessive heat. The thermal stress is the main cause for electrical machine failure. To prevent such failures in induction motor and to increase the machine reliability, various temperature monitoring methods are developed. The main concern in applying ML algorithm to SCIM is to predict temperature rise in stator and also in rotor. The standard specifications of 15 kW SCIM is collected and simulation is performed. The outer frame size of 15 kW is 160L. For simulation speed, current, voltage, frequency, torque, power factor, stator resistance, slip, loss resistance, efficiency is given as input. With the SCIM loss data, slip, current the heat source for each part of the SCIM is calculated. The heat source is taken as input for simulation. The temperature rise for all the parts is recorded. The obtained temperature rise data is verified with mathematical model. The same process is followed and applied for 160L frame series motors such as 18.5 kW, 15 kW, 11 kW, and 9.3 kW. Now with this data set from different rating motors the ML algorithm is applied to predict the temperature rise without performing any simulation. The temperature rise data set for different rating motor is given in [13, 14].

In this paper the prediction of temperature rise in SCIM due to variation in load is dis-cussed. In accordance with load variation, fault data is categorized into four groups. No -load for the first class, half load for the second, three-quarter load for the third, and full load for the fourth. The numerical designations for the classes are as follows: 1, 2, 3, and 4 loaded with the name 'Test data' and containing the integers 10, 1s through 10, 680 s, these classifications are imported into MATLAB for analysis purposes. The available test data comprises a total of 680 sets. For each class, ten data sets are evaluated.

3 Feature Classification and Extraction of Temperature Rise

Feature Extraction (FE): -This time segmentation is employed in feature extraction. Time Windowing (TW) was used as the segmenting the data set. The provided dataset is partitioned into distinct pieces. The temporal segmentation is utilized in feature extraction. The TW approach divides the data set into equal subsets with equal time periods. This strategy is utilized when the provided dataset is inundated. The primary drawback of this strategy is that a small dataset will be spread too thinly throughout the window, while a large dataset would be spread too widely. The selected methods for feature extraction are listed in Table 1. These feature extractions will yield precise findings in a shorter amount of time. The TW approach divides the data set into equal sets with equal time periods. This FE method is used when the specified data collection gets inundated [11].

Table 1. Categories of FE

Mean Absolute Value (MAV)	$\overline{X_i} = \sum_{k=1}^{N} \lvert x_k \rvert$, for i = 1,......... i-1 Where, N and i are the number of samples and channels
Integrated Absolute Value (IAV)	$IAV = MAV\text{-}N$ where N is number of samples
Root Mean Square (RMS)	$RMS = \sqrt{\frac{1}{n}\left(x_1^2 + x_2^2 + \ldots + x_n^2\right)}$ Where 'n' is the total number of samples
Waveform Length (WL)	$WL = \sum_{K=2}^{N} \lvert x_k - x_{k-1} \rvert$ N denotes the quantity of time samples, k is the samples and k-1 is previous samples
Zero Crossing (ZC)	$sgn(-x_k \times x_{k=1})$ $\lvert x_k - x_{k+1} \rvert \geq threshold$ If the threshold is exceeded and the *sgn* function returns a positive number, the zero crossing count is incremented

Feature Classification (FC):- ML encompasses several approaches for classifying features. These algorithm exhibit enhanced performance and rapid response times. The algorithms used in feature classification is listed in Table 2 [11]:

The process for present research work is shown in Fig. 2. From the flow chart it can be observed that, the 160 L Frame motor temperature rise simulation is performed and data is recorded. The raw data is recorded from simulation.

Table 2. Algorithms used in FC

Linear Discriminant Analysis (LDA)	LDA is a method that is frequently used to classify data. When there are differences in within-class frequencies and test data generated at random, LDA can handle this situation with ease
Quadratic Discriminant Analysis (QDA)	Each matrix covariance for a given set of data is computed as the variance of LDA. Information pertaining to a class data set is necessary for QDA to be beneficial
K-Nearest Neighbor (K-NN)	The algorithm is the most basic kind of ML and operates using the supervised learning methodology. K-NN operates on the assumption of similarity between a new fault data set and an existing fault data set that is readily accessible
Navi-Bayesian Pattern Classification (NBPC)	It is a ML algorithm that operates under the supervision of a human and is grounded in the principles of Bayes' theorem. Its purpose is to resolve classification issues. Its primary use is in the categorization of training data with a high number of dimensions
Decision Tree (DT)	A decision tree is a non-parametric supervised learning algorithm for classification and regression tasks. It has a hierarchical tree structure consisting of a root node, branches, internal nodes, and leaf nodes. Decision trees are used for classification and regression tasks, providing easy-to-understand models

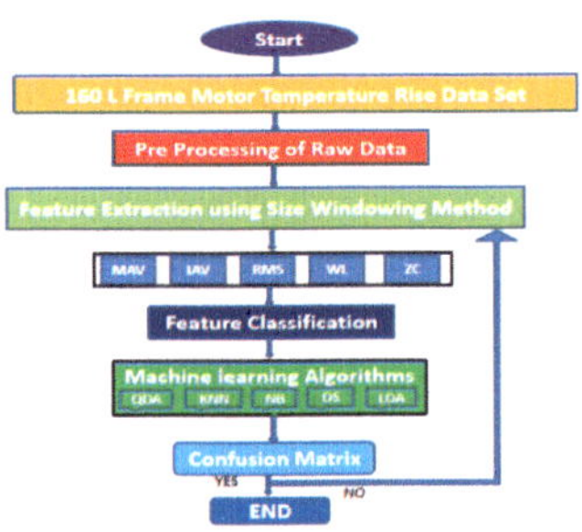

Fig. 2. Flow chart of prediction of temperature rise in SCIM

4 Results

This section discusses the research results. For the research 160 L frame based different rating SCIM is taken. The simulation is performed on 160 L frame based different rating SCIM to predict the temperature rise. The main focus of the simulation is to predict the temperature rise in 160 L frame based different rating SCIM with different loads. The current is varied for different loads and all parameters are recorded. Due the load is applied continuously, temperature rise is present in winding resistance. The simulation is performed on all 160 L frame based different rating SCIM: 18.5 kW, 15 kW, 11 kW, and 9.3 kW. The data sets are uploaded into the ML programme and separated into distinct classifiers, such as full load, 3/4th load, half load and no load. The Table 3 compares different feature classifications with regard to different feature extractions. A confusion matrix will provide a summary of the classification algorithm's properties. The performance of confusion matrix is computed as follows:

$$Performance = \frac{sum(diag(cmay))}{sum(cmat(:))} \times 100\% \tag{1}$$

Table 3. Temperature prediction performance utilizing FE and FC techniques

Feature Extraction	Feature Classification					
		KNN	LDA	DS	NB	QDA
	MAV	98%	95%	96%	95%	88%
	IAV	97%	88%	92%	91%	91%
	RMS	97%	98%	90%	96%	75%
	WL	96%	53%	90%	91%	95%
	ZC	72%	35%	50%	65%	70%

5 Conclusion

The ML system is designed to forecast temperature in 160 L frames with varied SCIM ratings of 18.5 kW, 15 kW, 11 kW, and 9.3 kW. This data is analysed using several feature extraction and classification approaches. LDA has been identified as the least effective feature classification technique, while K-NN has been identified as the most effective. For the purpose of predicting temperature rise, the performance of various feature extraction and classification methods is contrasted. An impressive 92% is achieved by the K-NN in this domain. Availing outcomes is achieved by employing the mean performance. But the time required for the procedure is longer in other feature categories. The K-NNA feature classification approach simplifies and accelerates the procedure. QDA scores 84%, NBPC scores 88%, DS scores 84%, and LDA scores 74% in terms of performance.

References

1. Czerwinski, D., Geca, J., Kolano, K.: Machine learning for sensorless temperature estimation of a BLDC motor. Sensors **21**(14), 4655 (2021)
2. Salinas, D.A.M.: Predictive Maintenance of Induction Motors using Deep Learning, degree project in technology. PhD Thesis. KTH Royal Inst. of Tech (2022)
3. Soprun, O., et al.: Forecasting temperatures of a synchronous motor with permanent magnets using machine learning. Creative Commons License Attribution 4.0 International (2020)
4. Guo, H., Ding, Q., Song, Y., Tang, H., Wang, L., Zhao, J.: Predicting temperature of permanent magnet synchronous motor based on deep neural network. Energies **13**(18), 4782 (2020)
5. Nogay, H.S.: Prediction of internal temperature in stator winding of three-phase induction motors with ANN, European transactions on electrical power. Eur. Int. Trans. Electr. Energy Syst. **21**, 120–128 (2011)
6. Dambrauskas, K., Vanagas, J., Zimnickas, T., Kalvaitis, A., Bizimavičius, T.: Induction motor's bearing condition motoring and diagnosis applying cloud services and artificial neural networks. IEEE 7th IEEE Workshop on Advances in Information, Electronic and Electrical Engineering (AIEEE), Liepaja, Latvia, pp. 1–5 (2019)
7. Singh, J., Singh, S.: Transformer failure analysis: Reasons and methods. Int. J. Eng. Res. Technol **4**(15), 1–5 (2016)
8. Park, B.-G., Lee, K.-J., Kim, R.-Y., Hyun, D.S.: Low-cost fault diagnosis algorithm for switch open-damage in BLDC motor drives. J. Power Electron. **10**, 702–708 (2010)
9. Guo, H., Ding, Q., Song, Y., Tang, H., Wang, L.: Predicting temperature of permanent magnet synchronous motor based on deep neural network. Energies **13**, 4782 (2020)
10. Sudha, B., Praveen, L.S., Vadde, A.: Classification of faults in distribution transformer using machine learning. Sci. Direct, Mater. Today Proc. **58**, 616–622 (2021)
11. Praveen, L.S., Nagananda, S.N., Shankapal, P.: Design and development of real time bionic hand control using EMG Signal. In: 2018 IEEE International Conference on Electronics, Computing and Communication Technologies, Bangalore, India, pp. 1–4 (2018)
12. Khazi, S., et al.: Analyzation of temperature rise in induction motor for electric vehicles. In: Bansal, R.C., Agarwal, A., Jadoun, V.K. (eds.) Advances in Energy Technology. LNEE, vol. 766, pp. 173–183. Springer, Singapore (2022). https://doi.org/10.1007/978-981-16-1476-7_17
13. Sudha, B., Vadde, A., Manickavasagam, K.: Thermal behavior on productive torque in electric vehicle motor using computational methods. In 2021 IEEE 6th International Conference on Computing, Communication and Automation (ICCCA), pp. 367–371 (2021)
14. Sudha, B., Vadde, A., Manickavasagam, K.: Characterization of temperature and productive torque for 160L frame squirrel cage induction motor. J. Eng. Res. (2021). EMSME Special Issue

Quantum Many-Body Problems: Quantum Machine Learning Applications

Pramoda Medisetty[1], Veda Manohara Sunanda Vulavalapudi[1], Poorna Chand Evuru[1], Mufti Mahmud[2], and Kolla Bhanu Prakash[1(✉)]

[1] Department of Computer Science and Engineering, Koneru Lakshmaiah Education Foundation, Vaddeswaram, Andhra Pradesh, India
drkbp@gmail.com

[2] Department of Computing and Technology, Nottingham Trent University, Nottingham, UK
mufti.mahmud@ntu.ac.uk

Abstract. A groundbreaking approach has been developed to accurately depict and predict the essential traits of quantum many-body systems. This inventive technique merges quantum ground state computations, classical shadow representations, and sophisticated machine learning approaches. The newly engineered software precisely identifies the ground state of a 2D antiferromagnetic Heisenberg model and forms a classical shadow representation for the quantum state. Machine learning models are deployed to forecast ground-state correlation functions with remarkable accuracy. This pioneering technique shows enormous potential in transforming quantum simulations and quantum machine learning, offering flexible solutions for complex quantum systems.Its implications extend to fields such as material science and drug discovery.

Keywords: Quantum Ground State · Quantum Many-Body Systems · Quantum Machine Learning · Quantum Simulation · Quantum Computational Tasks · Quantum Wavefunction

1 Introduction

The intricacies of quantum many-body systems have long challenged scientific exploration due to their complex behaviors governed by the laws of quantum mechanics. These systems hold the promise of revolutionizing various fields, from condensed matter physics to quantum computing. Yet, efficiently representing and predicting the ground-state properties of these systems remains a formidable task. This research introduces an innovative methodology that harnesses classical shadow representations and machine learning techniques to predict ground-state properties for quantum many-body systems. The focal point of this study is the 2D antiferromagnetic Heisenberg model on a square lattice, a quintessential quantum system. The methodology comprises key steps. First, a Hamiltonian defining quantum interactions within the system is precisely determined. The ground state is then calculated using SciPy, providing the foundation for subsequent analysis. The research evaluates correlation functions for pairs of quantum spins

V. E. Balas et al. (Eds.): IEQIP 2023, LNNS 1029, pp. 79–86, 2024.
https://doi.org/10.1007/978-3-031-61929-8_12

using Pauli operators and constructs a correlation matrix C. The groundbreaking concept of classical shadows emerges, representing the quantum state with high efficiency through randomized single-qubit measurements. From these classical shadow snapshots, an approximate quantum state is reconstructed. Classical machine learning models like Support Vector Regression (SVR) and Kernel Ridge Regression are employed to predict ground-state properties, notably the complex correlation functions. The application of classical machine learning models, specifically Support Vector Regression (SVR) and Kernel Ridge Regression, to predict correlation functions in quantum systems represents a sophisticated and powerful approach with significant implications for understanding complex physical phenomena. In this context, these models are skillfully trained to extract meaningful information from data sets associated with quantum many-body systems, particularly focusing on correlation functions. Support Vector Regression (SVR) and Kernel Ridge Regression are both regression techniques that excel in capturing complex relationships within data. In the realm of quantum physics, correlation functions play a pivotal role in characterizing the behavior of quantum particles and their interactions. These functions provide crucial insights into the correlations between quantum observables, offering a window into the intricate dynamics of quantum many-body systems. The skillful training of SVR and Kernel Ridge Regression involves exposing these models to labeled data sets containing information about the quantum system's correlation functions. The models learn to generalize from this data, enabling them to make accurate predictions for correlation functions in scenarios where the ground-state properties of the quantum system are not explicitly known. One of the strengths of Support Vector Regression lies in its ability to handle non-linear relationships effectively. This is particularly advantageous in the quantum realm, where complex interactions and correlations often exhibit non-linear behavior. SVR aims to find a hyperplane that best captures the underlying patterns in the data, allowing for accurate predictions of correlation functions even in highly non-linear quantum systems.

Similarly, Kernel Ridge Regression leverages the concept of kernels to map input data into a high-dimensional space, where linear regression is applied. This technique is adept at capturing complex relationships and is well-suited for scenarios where the correlation functions exhibit intricate patterns and dependencies. The incorporation of hyperparameter tuning further enhances the predictive capabilities of these models. Hyperparameter tuning involves optimizing the parameters that govern the learning process, ensuring that the models are finely tuned to the specific characteristics of the quantum system under investigation. This iterative refinement process contributes to the accuracy and reliability of the predictions, making the models more adept at capturing the nuances of quantum correlation functions. By employing classical machine learning models like SVR and Kernel Ridge Regression, researchers can extract valuable insights into the correlation functions of quantum many-body systems. The ability to predict these functions not only facilitates a deeper understanding of the quantum dynamics at play but also opens avenues for exploring and manipulating quantum states for various applications in fields such as quantum computing, materials science, and condensed matter physics. Overall, the skillful application of classical machine learning techniques in predicting correlation functions represents a promising and innovative approach to unraveling the complexities of quantum systems. Hyperparameter tuning further refines the machine learning models. Each correlation function becomes a gateway to understanding quantum properties, and the models are evaluated to determine the best fit for each. It bridges the quantum-classical divide, offering a methodology combining classical shadows and machine learning for predicting ground-state properties.

2 Literature Survey

Quantum computing circuits and devices have seen significant development in recent years, driving advancements in quantum technology [1]. A comprehensive resource on quantum computation tools and specialized topics has been instrumental in shaping the field [2]. The application of supervised learning with quantum computers has gained substantial attention in quantum computing research [3]. Pioneered in 1996, Grover's quantum algorithm for database search stands as a foundational milestone in quantum computation [4]. The role of quantum computing circuits and devices in advancing quantum technology is emphasized in current research [1]. A detailed survey on quantum computing technology offers insights into the state of the field and potential applications [5]. Quantum computing has shown promise in innovative approaches to tasks such as image classification through the use of algorithms like the quantum K-Nearest-Neighbor algorithm [6]. Quantum algorithms, such as algorithms based on Hamming distance metrics, have opened new avenues for data analysis and classification [7].

3 Methodology

It is an innovative approach to predict the ground-state properties of quantum many-body systems with remarkable efficiency. The investigation focuses on the 2D antiferromagnetic Heisenberg model on a square lattice as the quantum system of interest. At the outset, the Hamiltonian accurately delineates how spins interact within the quantum system, forming the theoretical groundwork for upcoming computations. An essential element of this approach is the creation of the J_mat coupling matrix, which adeptly identifies the complex web of links quantum spins, as depicted in Fig. 1.

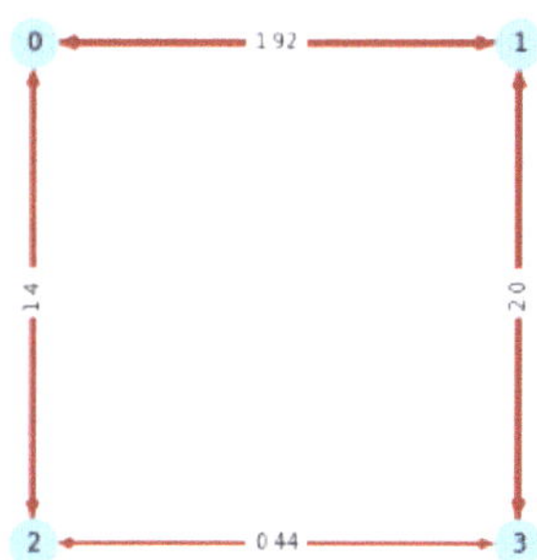

Fig. 1. Coupling matrix J_mat

The classical shadows offer an approximation of the quantum state with enhanced efficiency and reduced storage requirements. To facilitate comprehension a network diagram is created that visually illustrates how the components of the quantum system are connected. The susequent critical step centers on identifying the system's ground state. To accomplish this, we harness the robust SciPy library to test the Hamiltonian, enabling us to calculate the eigenvalues and eigenvectors of the ground state. Once the ground state is established, the attention switches to the computation of the correlation matrix. This body includes two-body correlation methods, named as C_ij, stated in terms of Pauli operators, as shown in Fig. 2. The correlation matrix, referred to as C, is then methodically stated, explaining the characteristics of the quantum system. Breaking away from classical outcomes of quantum states, this approach introduces classical shadow representations. These outcomes are build through randomized measurements of single-qubits across multiple instances of the quantum state.

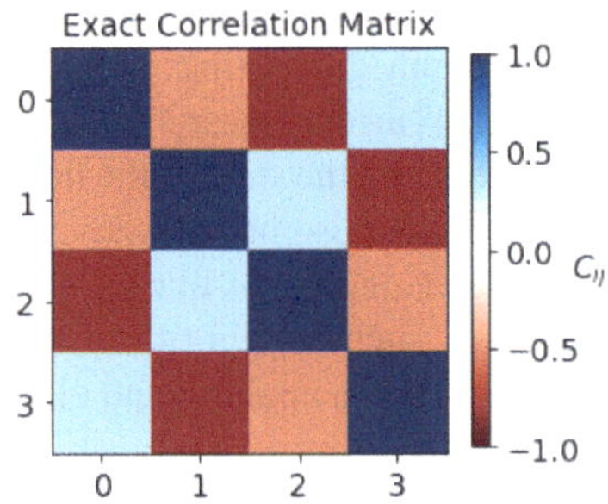

Fig. 2. Exact Correlation Matrix

Building upon these classical shadows, the methodology proceeds to reconstruct an approximate quantum state. Each snapshot of the classical shadow undergoes a meticulous process to produce an estimation of the quantum state. The estimation of the correlation matrix, denoted as C', from the classical shadow approximation of the ground state is carried out. Robust estimation methods, including median of means estimation, are employed to calculate expectation values of the correlation functions based on the reconstructed quantum state. A significant aspect of this methodology is the application of classical machine learning models, including Support Vector Regression (SVR) and Kernel Ridge Regression. These models are trained with a dataset comprising coupling vectors and estimated correlation matrices. Hyperparameter tuning is meticulously conducted to optimize the machine learning models for predicting correlation functions. This process involves iterating through different regularization parameters and models to identify the most fitting configurations. With the machine learning models primed and ready, the methodology proceeds to predict ground-state properties. For each correlation function C_ij, ground-state properties are predicted utilizing the trained models. Model performance is evaluated, enabling the identification of the optimal model for each C_ij as shown in Fig. 3. In the final stages of the methodology, a comprehensive model evaluation is performed by comparing the predictions of the machine learning models to the exact ground-state correlation matrices. This critical evaluation assesses the accuracy and effectiveness of the models in approximating quantum properties. It's harnesses classical shadow representations and classical machine learning models to predict ground-state properties of quantum many-body systems with remarkable efficiency. It seamlessly integrates theoretical groundwork, computational calculations, classical approximations, and machine learning to offer a systematic and practical approach for tackling complex quantum systems.

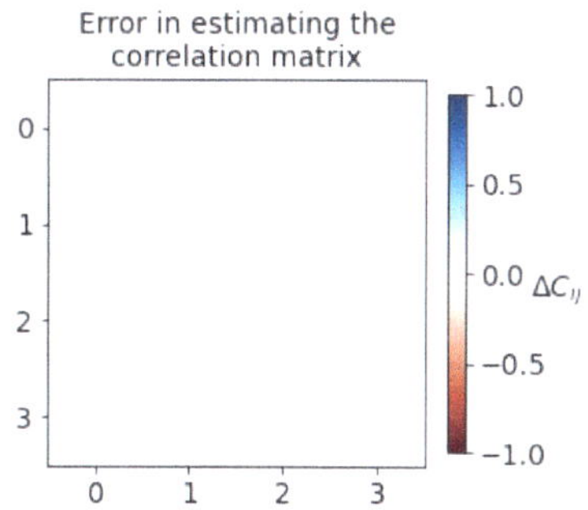

Fig. 3. Error in estimating the correlation matrix

4 Results and Discussion

It showcases a comprehensive and successful methodology for efficiently predicting the ground-state properties of quantum many-body systems, with the specific example of the 2D antiferromagnetic Heisenberg model on a square lattice. The quantum ground state calculation, facilitated by SciPy's eigenvalue solver, yielded precise ground state energies and eigenstates. The computed quantum wavefunction allowed for the construction of the correlation matrix C, shedding light on the intricate relationships between quantum spins. A pivotal aspect of this research is the introduction of classical shadow representations, enabled by randomized single-qubit measurements. These classical shadow snapshots not only efficiently captured the quantum state but also significantly reduced storage requirements, presenting an innovative solution for representing complex quantum systems classically.

The research further demonstrated the application of classical machine learning models, such as Support Vector Regression (SVR) and Kernel Ridge Regression, which are skilfully trained to predict correlation functions. An intricate hyperparameter tuning process optimized these models, revealing the best regularization parameters for accurate correlation function predictions. The predictive performance of the trained machine learning models was outstanding, successfully forecasting the ground-state correlation functions for various pairs of quantum spins. Rigorous model evaluation allowed for the identification of the most suitable models for each correlation function, underlining the potential of this methodology to efficiently address complex quantum systems and significantly impact the fields of quantum simulations and quantum machine learning.

5 Conclusion

In conclusion, our research has unveiled a promising and innovative methodology that brings us one step closer to predicting the ground-state properties of quantum many-body systems. The central findings emphasize two critical aspects. Firstly, the introduction of classical shadows as an efficient representation of quantum states has the potential to revolutionize the field. The reduced computational and storage demands associated with this approach open new horizons for scalable quantum simulations, eliminating the burdens typically associated with managing full quantum states. Secondly, the successful application of classical machine learning models, meticulously trained on data derived from classical shadows, underscores their ability to accurately forecast ground-state properties, particularly in the context of correlation functions for quantum systems. The models demonstrated a remarkable predictive accuracy, effectively capturing the intricate quantum correlations and properties within the confines of the 2D antiferromagnetic Heisenberg model.

The application of classical machine learning models, specifically Support Vector Regression (SVR) and Kernel Ridge Regression, has yielded remarkable results in predicting correlation functions within the context of the 2D antiferromagnetic Heisenberg model. These models showcase an exceptional level of predictive accuracy, effectively capturing the intricate quantum correlations and properties inherent in this fundamental quantum system. The 2D antiferromagnetic Heisenberg model on a square lattice represents a challenging quantum system with complex interactions between quantum spins. Understanding and predicting the correlation functions in such systems are crucial for unravelling the underlying quantum dynamics and gaining insights into the material's behavior. The models, having been meticulously trained on relevant data sets derived from the quantum interactions within the Heisenberg model, exhibit a remarkable ability to generalize and predict correlation functions accurately. This predictive accuracy is particularly noteworthy given the inherently intricate nature of quantum correlations, which often involve non-linear and complex relationships between observables. The success of the models in capturing the quantum correlations suggests that classical machine learning techniques, when skillfully applied, can effectively bridge the gap between classical and quantum domains. The ability to predict correlation functions in the 2D antiferromagnetic Heisenberg model opens avenues for exploring emergent quantum phenomena, guiding experimental investigations, and advancing our understanding of condensed matter physics. Furthermore, the term "intricate quantum correlations and properties" emphasizes the nuanced and complex nature of the quantum interactions involved. The models demonstrate an aptitude for not only reproducing known correlations but also uncovering subtle relationships and properties that contribute to the overall richness of the quantum system. This achievement is not only a testament to the power of classical machine learning in handling quantum complexities but also signifies a significant step toward a unified approach in tackling quantum many-body systems. The models' success in capturing the intricacies of the 2D antiferromagnetic Heisenberg model showcases their potential applicability in broader contexts, promising advancements in quantum computing, materials science, and other quantum-related fields. In summary, the application of classical machine learning models, specifically SVR and Kernel Ridge Regression, has proven to be highly successful in predicting correlation functions within the 2D antiferromagnetic Heisenberg model. The models' remarkable predictive accuracy underscores their potential in unraveling the intricate quantum correlations and properties of complex quantum systems, marking a substantial advancement in the intersection of classical machine learning and quantum physics.

While these findings offer great promise, there are still considerations and avenues for future exploration. Sensitivity to measurement errors is an inherent limitation of the classical shadow approach, and future research endeavours may focus on enhancing its robustness, ensuring reliability in the face of real-world imperfections. Moreover, the broad-reaching implications of this research cannot be understated. The demonstrated methodology holds immense significance in the development of quantum simulations and quantum machine learning algorithms. These advancements not only offer solutions to the inherent challenges of quantum many-body physics but also extend to a wide array of practical applications in material science, drug discovery, and other quantum computational tasks, thereby transcending the boundaries of fundamental research and venturing into the realm of tangible real-world implications. In essence, the code and methodology presented in this study contribute significantly to the ongoing exploration of efficient classical representations of quantum states. They offer the potential to revolutionize quantum simulations by providing practical, scalable solutions for complex quantum systems. As the scientific community continues to push the boundaries of quantum research, this methodology stands as a beacon of innovation and an exemplar of the path forward in the quest for efficient quantum solutions as shown in Fig. 4.

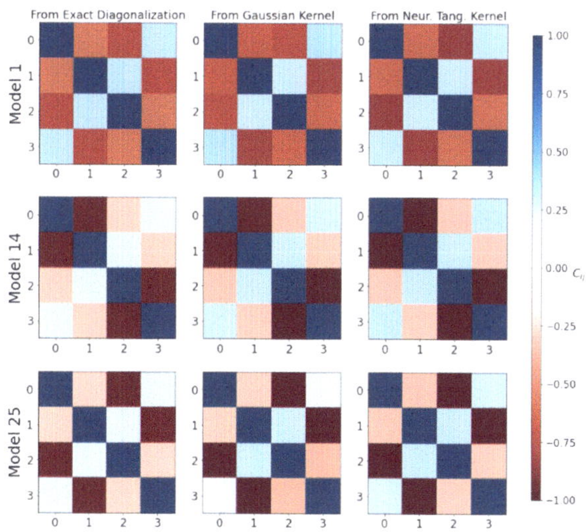

Fig. 4. Correlation Matrix for Different Kernals

References

1. Humble, T.S., Thapliyal, H., Munoz-Coreas, E., Mohiyaddin, F.A., Bennink, R.S.: Quantum computing circuits and devices. IEEE Des. Test. **36**(3), 69–94 (2019)

2. Benenti, G., Casati, G., Strini, G.: Principles of Quantum Computation and Information-Volume II: Basic Tools and Special Topics. World Scientific Publishing Company (2007)
3. Schuld, M., Petruccione, F.: Supervised Learning with Quantum Computers. Springer, Berlin (2018)
4. Grover, L.K.: A fast quantum mechanical algorithm for database search. In: Proceedings of the Twenty-Eighth Annual ACM Symposium on Theory of Computing, pp. 212–219 (1996)
5. Gyongyosi, L., Imre, S.: A survey on quantum computing technology. Comput. Sci. Rev. **1**(31), 51–71 (2019)
6. Dang, Y., Jiang, N., Hu, H., Ji, Z., Zhang, W.: Image classification based on quantum K-Nearest-Neighbor algorithm. Quantum Inf. Process. **17**(9), 1–8 (2018)
7. Ruan, Y., Xue, X., Liu, H., Tan, J., Li, X.: Quantum algorithm for k-nearest neighbors classification based on the metric of Hamming distance. Int. J. Theor. Phys. **56**(11), 3496–3507 (2017)

Experimental Study on the Impact of Airborne Dust Deposition on PV Modules Using Internet of Things

Gowtham Vedulla and A. Geetha(✉)

Department of Electrical and Electronics Engineering, Faculty of Engineering and Technology, SRM Institute of Science and Technology, Kattankulathur Campus, Chennai, TN 603203, India
geethaa2@srmist.edu.in

Abstract. The increasing demand for Various energy sources have emerged as a result of energy and alternatives, but PV systems have shown to be the most efficient. This level of reliance on solar power is obvious given with reliable power supply and minimal adverse effects. There are still other problems that have an adverse effect on the operation of PV systems, including as temperature-induced voltage drops that cause solar power to be lost and dust buildup (soiling). Dust accumulation has a major impact on the optical properties of PV modules, which lowers the output of generated photocurrent. The performance of solar cells is influenced by both inner and external considerations, including location, manufacturing processes, wind, dust, humidity, and other environmental factors. The efficiency of solar systems is influenced by a variety of elements, including temperature, ageing, wind, radiation, pollution, shade, and cleanliness. Dust is a type of minute particles that are released into the air by building sites, windstorms, and industry pollutants. Either it is visible or it is not. This project's goal is to employ thing speak web server to analyzed how dust affects the performance of the PV system at SRM University in Chennai. The characteristic curve was used to compare all the significant characteristics (output power, short circuit current, open circuit voltage, etc.) relevant to the two panels. After 45 days, there was a 7.5% drop in the soiled panel's ability to generate electricity.

Keywords: Dust impact · Solar Photovoltaic · System Performance · Thingspeak web server

1 Introduction

At this point, when natural energy resources are being depleted at a rapid rate, this issue makes it necessary for the globe to use all available resources as effectively as possible. Although it has not been fully utilized, the finest source of energy creation in this situation has emerged as solar energy. This can be accomplished by creating PV cells that are more effective and enhanced as well as by minimizing losses throughout the production and transmission processes. The buildup of dust on the Solar panels surface, that results in a significant energy loss during long-term manufacturing, is one of the main issues hurting the production of PV modules [1].

V. E. Balas et al. (Eds.): IEQIP 2023, LNNS 1029, pp. 87–93, 2024.
https://doi.org/10.1007/978-3-031-61929-8_13

Multiple ecological conditions, including wind velocity, radiation from the sun, ambient temperature, humidity, and dust, have an impact on performance of Photovoltaic modules [2]. The ambient temperature of a solar cell has a significant impact on how much power it produces. Most photovoltaic cells have been standardized to work at a standard temperature of 25 °C and so the performance of the solar module declines as the outside temperature rises [3]. The performance of solar cells is significantly impacted by wind speed in a different way [4]. The efficiency of the cell improves when wind velocity increases because it lowers the cell's ambient temperature. Additionally, it aids in clearing the cell's surface of larger dust particles. Many photons can reach the surface when these particles are removed, increasing power generation.

The modulation of the light path brought on by water drops that are present in the surroundings as moisture particles is one of two ways that humidity affects a PV cell's performance. Low performance is brought on by this reduction in the direct sun radiation component's reception level. The water particles are distributed around the cell has another effect as a result of humidity. Long-term exposure to higher humidity levels close to the cell can harm it and lower both performance and life span [5]. As can be observed, solar panels are typically positioned in places where there shouldn't be any barriers between the sun's rays and the panels, allowing the wind to freely disperse dust. Dust deposition is a fairly complicated process that randomly reduces performance based on the location and other variables [6]. The term "dust" refers to tiny solid particles with a diameter smaller than 500 m. When they settle on a surface, microscopic pollens like germs and fungus, as well as microfibers removed from clothing, fabrics and carpets are also regarded as dust particles [7]. Depending on a number of variables, the distribution of the dust particles that had collected on the panel may or may not be uniform.

The power density of sunlight at a surface per unit area in the form of solar radiation is known as solar irradiance and is expressed in watts per meter square [8]. Performance of solar cell increases proportionally with irradiance intensity maximum power, photocurrent and Short circuit current grow directly as irradiance intensity rises, but voltage and fill factor only somewhat so [9]. The buildup of dust on solar module surfaces is a significant problem, particularly in desert areas where dust storm frequency is quite high. There haven't been many studies on how dust affects PV cell performance in arid environments [10]. The majority of study on how dust accumulation affects the performance of solar cells have either been conducted indoors or in locations other than deserts [11]. The local environment, inclination angle, and All solar panel surfaces have a significant impact on the accumulation of dust [12]. The term "native environment" describes the surrounds that human activity has directly or indirectly produced, such as the built environment, the kinds of vegetation, and the weather [13]. Heavy dust can gather on a surface the more horizontal it is. While a powerful wind might clean the panel surface, a slow breeze can also cause dust to accumulate. The solar panel's dust collector greatly benefits from the surface as well. More soil can accumulate If the surface is not smooth but is rough, rage, sticky, etc.

2 Reasons for Dust Implementation

The local environment, inclination angle, and All solar panel surfaces have a significant impact on the accumulation of dust [12]. The term "native environment" describes the surrounds that human activity has directly or indirectly produced, such as the built environment, the kinds of vegetation, and the weather [13]. Heavy dust can gather on a surface the more horizontal it is. While a powerful wind might clean the panel surface, a slow breeze can also cause dust to accumulate. The solar panel's dust collector greatly benefits from the surface as well. More soil can accumulate If the surface is not smooth but is rough, rage, sticky, etc. [14].

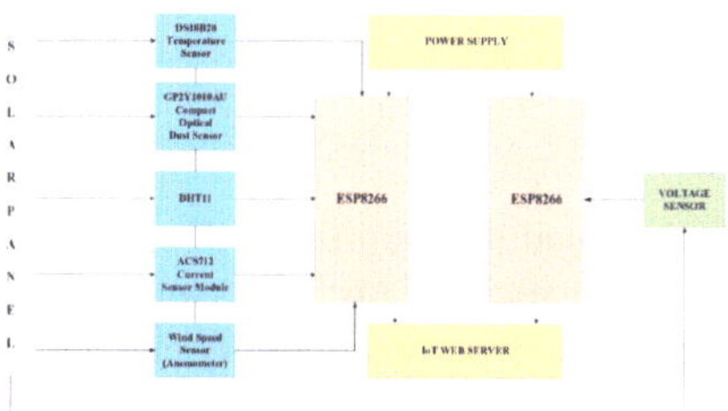

Fig. 1. Block Diagram of solar efficiency monitoring

Fig. 2. Solar Module

As seen in Fig. 1, Sensor translates the physical quantity it perceives into a measurable electrical characteristic. DHT11 for read environmental temperature and humidity values. ESP8266 has inbuilt wifi module and reads sensor data and send the sensor data to Thingspeak web server. Thingspeak web server is Iot platform stories all data time to time.

3 Design of an Experimental Setup

Figure 2 display Solar pv module(40W) and The objective of our research is to compare the effects of dust on these two polycrystalline photovoltaic panels under identical conditions. For our project, two straightforward requirements had to be met while choosing a site. Which are listed below (Table 1).

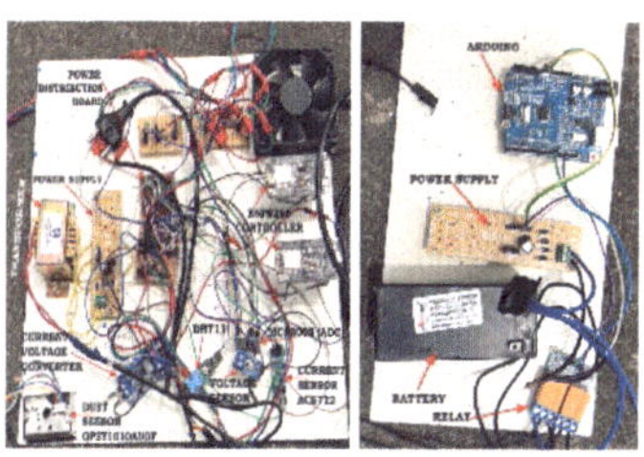

Fig. 3. Components of Experimental Setup

Fig. 4. Surface of solar Panel after 45days

- No whole day shadowing allowed
- Simple wiring for our test setup

Table 1. Shows the most important aspects of solar modules.

Specifications	Poly-crystalline
Maximum rated power (Pmax)	40 W
Operating rated voltage (Vmax)	17.40 V
Operating rated current(Imax)	2.30 A
Open circuit voltage (Voc)	21.90 V
Short circuit current (Isc)	2.45 A

In order to examine the effects of external conditions on the performance of the photovoltaic system, Fig. 3 displays experimental setup of installed two photovoltaic panels with similar specifications in the same location on the university campus.

4 Result Analysis

As is seen in Fig. 4, we examined and reported on the PV system's performance as a result of dust buildup during a 45-day period. This report includes a graphic comparison and record of many variables. Both panels were placed in a similar environment for 45 days, during which time one panel received routine maintenance while the other one was abandoned without cleaning.

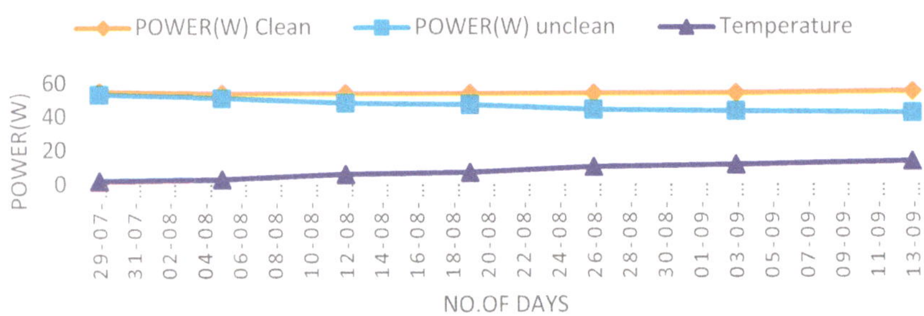

Fig. 5. Shows an output power comparison between clean and non-clean solar panels.

The values of all relevant parameters from both panels were recorded by the Thingspeak web server who then stored the information in an excel file. The P-V curve, P-t curve and I-V curve for both pure and soiled solar panels were established over a 45-day period. As is seen in Fig. 5, the power curves of the two panels are practically identical at first, but as time goes on, the output power of the panels gradually differs from one another. We can easily see that the difference percentage increases to 7.5% since one panel had dust buildup while the other had frequent cleanings, so its performance was almost constant.

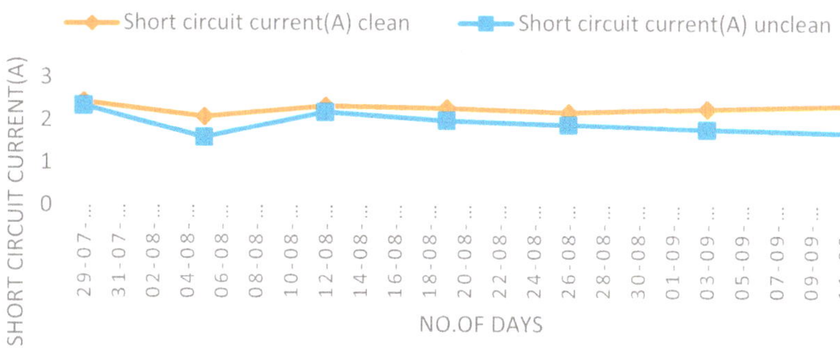

Fig. 6. Clean and dirty solar panel short circuit current curves

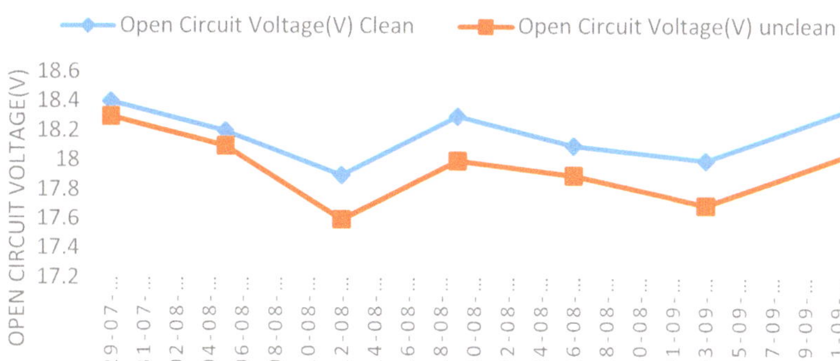

Fig. 7. Clean and dirty solar panel open circuit voltage curves

Figure 6 displays the current output of the uncleaned panel starts to decline significantly with time in contrast to the clean panel. It was crystal evident that as time went on, the surface's ability to absorb light was hampered by dust degradation.

We can infer from the preceding figures and the Fig. 7 above that the unclean panel performs worse than a pure one, however we can also come to this conclusion from this that when dust accumulation grows over time, the short circuit current reduces by a greater percentage than the open circuit voltage decreases.

5 Conclusion

This research objective was to learn more about how dust deposition effects PV module performance in desert environments. Over the course of 45 days. Two panels performance measures (one pure, the other not) were gathered. While two solar panels were subjected to the same conditions, only one was regularly cleaned during the course of the study. The panel's performance was examined by producing various characteristic curves. We were capable of understanding the occurrence of power loss as a result of dust buildup on solar panels cells due to the comparative analysis of the I-V curve. When the data and characteristics curve were analyzed, the solar cells open circuit voltage changed very slightly. However, due to dust accumulation, short circuit current was significantly impacted. Power output decreases as a result of this decrease in output current. After 45 days of data analysis, the unclean installed module's power output was 7.5% lower on day 45 compared to the cleaned one. A few of the dust particles that are accumulated on the surface cause the PV cell irreparable damage. We suggest that Under comparable desert circumstances, using a weekly cleaning schedule can help to increase the PV system's efficiency.

References

1. El-Shobokshy, M.S., Hussein, FM.: Degradation of photovoltaic cell performance due to dust deposition on to its surface. Renew Energy **3**(6/7), 585–90 (1993)
2. Hassan, A.H., Rahoma, U.A., Elminir, H.K., Fathy, A.M.: Effect of airborne dust concentration on the performance of PV modules. J Astron Soc Egypt **13**(1), 24–38 (2005)
3. Shaharin, AS., Haizatul, HH., Nik Siti, HNL.: Effect of dust on the performance of PV panels. World Acad. Sci. Eng. Techno. **58** (2011)
4. Neil, S., Robert, S.: Understanding the effect of sand and dust accumulation on photovoltaic modules. Renew. Energy **48**, 448–452 (2012)
5. Mekhilef, S., Saidur, R., Kamalisarvestani, M.: Effect of dust, humidity and air velocity on efficiency of photovoltaic cells. Renew. Sustain. Energy Rev. **16**, 2920–2925 (2012)
6. Shaharin, A.S., Atul, K.S., Mior, M.M.M.: Influence of dirt accumulation on performance of PV panels. Energy Procedia **50**, 40–56 (2014)
7. Mejia, F., Kleissl Bosch, J.L.: The effect of dust on solar photovoltaic systems. Energy Procedia **49**, 2370–2376 (2014)
8. Abhishek, R., Rohit, P., Monto, M., Praveen, R.: Influence of dust deposition on photovoltaic panel performance. Energy Procedia **54**, 690–700 (2014)
9. Shaharin, A.S., Mohamad, N.H.M.: Real-time study on the effect of dust accumulation on the performance of solar PV panels in Malaysia, In: Conference on Electrical & Information Technology, pp. 269–274 (2015)
10. Motasen, S., Abdul, G.A., Emil, A., John, K.K.: Experiment study on the effect of dust deposition on solar photovoltaic panels in desert environment. Renew. Energy **92**, 499–505 (2016)
11. Mohamed, Z., Abdullah, A.H., Ihab, E.S.: LabVIEW based monitoring system applied for PV power Station. In: International Conference on Automatic Control, Modelling & Simulation, ACMOS (2010)

12. Yousry, A., Mohamed, B.A.Z., Abdullah, A.H.: Solar cell curves measurement based on LabVIEW microcontroller interfacing. In: international conference on Automatic control, modelling & simulation, ACMOS (2010)
13. Gupta, V., Raj, P., Yadav, A., Garg, P., Nizam, F.: Design and analysis of PV analyzer on LABVIEW. In: 2nd international Conference on intelligent Communication, Control and Devices, ICICCD (2017)
14. Gupta, V., Raj, P., Yadav, A.: Investigate the effect of dust deposition on the performance of solar PV module using LABVIEW based data logger. In: IEEE International Conference on Power, Control, Signals and Instrumentation Engineering, ICPCSI (2017)

Bidirectional Converter with Time Utilization-Based Tariff Investigation and IoT Monitoring of Charging Parameters Based on G2V and V2G Operations

A. Geetha[1], S. Usha[1](✉), P. Geetha[2], and C. Jenifer Pon Catherin[2]

[1] Department of Electrical and Electronics Engineering, College of Engineering and Technology, SRM Institute of Science and Technology, Kattankulathur, Chennai 603203, India
ushas@srmist.edu.in

[2] Department of Electronics and Communication Engineering, Karpaga Vinayaga College of Engineering and Technology, Chennai 603308, India

Abstract. Load demand has been greatly impacted by the present boom in electric vehicle (EV) adoption, and this trend is expected to continue. As a result, the grid is under stress, and the recharging stations need to use effective energy management. Furthermore, new developments enable intelligent communication between an owner of an electric vehicle and a charging infrastructure. This infrastructure not only supports the grid when needed but can also charge the batteries of electric vehicles in both G2V and V2G modes. This article provides an IoT-based power scheduling for G2V and V2G modes of vehicle operation. It uses the time-based energy-based tariff to calculate costs in the G2V mode. On the other hand, V2G mode cost, which varies between ranges, is set through monitoring of the charging parameters of EV. Consider an hour of power usage with a 3-kW recharge based on a probabilistic model. The proposed model's information is saved on the cloud, and the data can then be retrieved through an online application for proper scheduling and pricing.

Keywords: G2V · V2G · Electric Vehicle · Internet of Things (IoT) · Cloud-based monitoring system

1 Introduction

During the initial period of significant development and adoption of electric vehicles in the early 20th century, commonly referred to as the first golden era of electric vehicles, there was a notable increase in the need for charging infrastructure to accommodate the growing demand for electric vehicle charging [1]. The initial standardization effort for electric vehicles centered on the development of the charging plug. This crucial component was initially introduced in 1913 and later gained international recognition as British Standard 74 following its approval. The aforementioned standard served as the inaugural benchmark for electric vehicles. The vehicle's design was additionally

V. E. Balas et al. (Eds.): IEQIP 2023, LNNS 1029, pp. 94–99, 2024.
https://doi.org/10.1007/978-3-031-61929-8_14

affected by considerations regarding the infrastructure. For instance, the selection of common battery voltages was made in order to facilitate direct charging from a 110 V DC power source. Electric vehicles provide a heightened level of adaptability in their potential uses due to their capacity to be equipped with onboard chargers and directly linked to the alternating current (AC) distribution network [2, 3].

V2G technology enables the transmission of surplus energy stored in the battery of an electric car back to the power grid, hence providing advantageous outcomes for the grid. This approach highlights the principle of utilizing electric vehicles and hybrids in a reciprocal manner, wherein the vehicles are connected to the grid for recharging purposes, and any surplus electricity is subsequently fed back into the grid [4]. Individuals who possess automobiles equipped with Vehicle-to-Grid (V2G) technology possess the capability to vend surplus electric power to power engineering experts during periods of vehicle inactivity. Subsequently, these individuals can utilize the proceeds from these transactions to recharge their vehicles during periods when the cost of electric power is comparatively lower [5].

In recent years, there has been a discernible inclination towards the integration of electric EVs inside intelligent urban environments. The reason for this is that EVs enable the facilitation of municipal carbon dioxide (CO_2) reduction initiatives [6]. Few studies have been proposed in the literature to examine strategies for enhancing the assimilation of EVs into mainstream society, with a focus on enhancing their adaptability and efficacy. A study was conducted utilizing an initiative architecture strategy to facilitate the digitalization of EVs in order to promote sustainable mobility in the context of electromobility. The objective of the work was to integrate data from several consumers and models pertaining to urban mobility systems [7, 8].

Entities like charging points and grid system operatives are integral components of the EV ecosystem [9]. The integration of smartness in charging and V2G charging techniques involves the establishment of a connection between an EV and the electrical grid through the utilization of a charge point equipped with information collection [10]. This facilitates the transfer of information and control signals among the different units within the EV ecosystem. When making decisions regarding the placement of charging infrastructures, urban areas are considering the level of readiness of their current infrastructure to accommodate the integration of vehicle-to-grid technology [11, 12].

The rest sections of this article are structured as: Sect. 2 presents a complete exposition of the proposed framework developed in the study. Section 3 delves into a comprehensive examination of the obtained results and subsequent analysis. Finally, the work is concluded in Sect. 4.

2 Proposed System Design

V2G represents a highly advanced and comprehensive approach, classified as level 5, within the realm of the IoT. This solution facilitates the seamless exchange of information within the electric grid among several users, hence significantly augmenting connectivity across various infrastructures. The technology is a management system that is based on the intelligence. The term "cloud storage" refers to the practice of storing and transferring data through an Internet gateway. In order for the V2G system to operate well, it is important to have both comprehensive engagement and a centralized method for coordinating the scheduling of EV activities during the charging and discharging modes as depicted in Fig. 1.

The proposed network management system for V2G is an IoT solution. The actuating signals originate from a single coordinator node, while many end nodes or client equipment transmit and receive information towards and from this node. The terminal nodes are responsible for transmitting data to the central node, which aggregates and subsequently transmits it to the cloud for further processing and storage in a database. Users have the ability to select their preferred method of accessing the cloud-hosted database, enabling them to engage in its operation in a manner that aligns with their individual requirements.

In order to gain access to the cloud, a coordinator node functions as a central access point to which a group of client devices establish connections. These devices are utilized to quantify the battery of a vehicle as a component of a battery management system, subsequently transmitting the outcomes wirelessly to the coordinator nodes that correspond with the data. The real-time calculation of the battery's SOC fluctuation is the duty assigned to the coordinator node. This approach aids in reducing the computing burden on the cloud infrastructure and mitigating latency.

During the process of scheduling V2G operations, the cloud server transmits the calculated variation in SoC and customer information to a cloud platform. The cloud platform utilizes a comprehensive database of customers, EVs, power grids, and charging infrastructures to provide optimal charging schedules for all EVs, taking into consideration variations in power pricing. The actuation signals specific to the device, such as charging and discharging, are transmitted to the coordinator node, which subsequently relays them to the electric cars under its supervision.

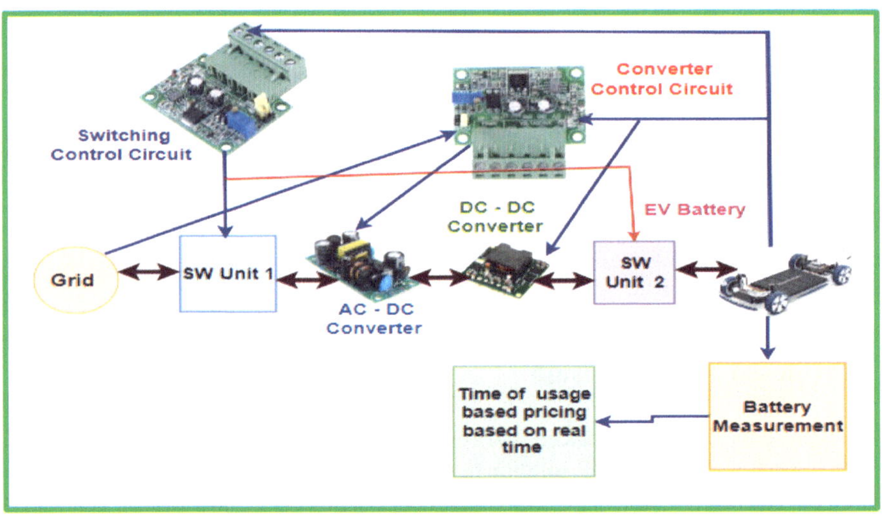

Fig. 1. Layout of proposed system

The growth of IoT systems is significantly impacted by the presence of security issues, which are of utmost importance. The cyber-physical networking model necessitates the inclusion of several components such as powerful access control, unspecified signatures, remote verification, data secrecy, message integrity, anonymous authentication, and a V2G network payment mechanism.

3 Results and Discussion

To transfer information needed the web cloud, an IoT device that communicates output is used. Think Speak is an entirely free cloud service that comprises Simulink and offers eight unlimited information transport channels. It is capable of updating within a short period. Battery charging state, energy sources, cost analysis, and time for parking are all sent to the think speak unit.

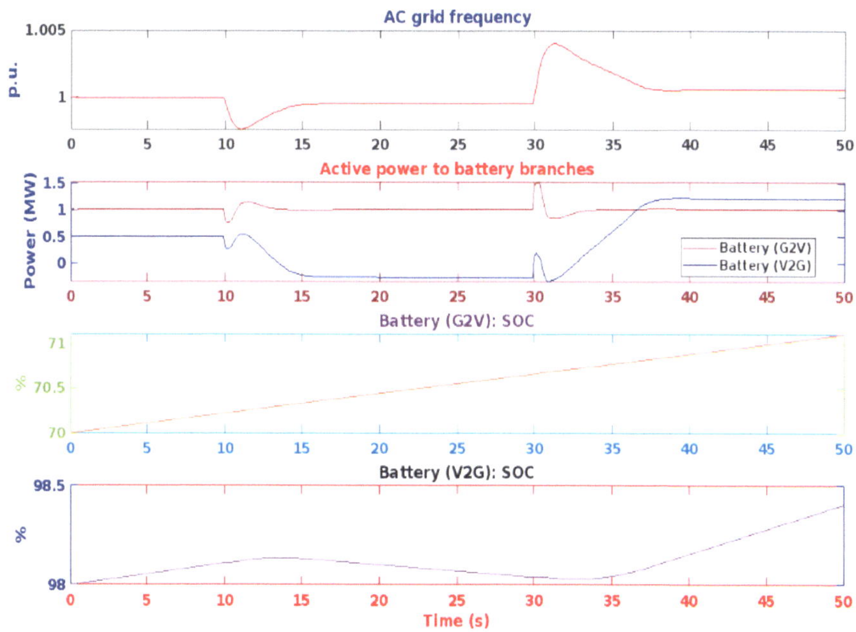

Fig. 2. Graphical Representation of AC grid Frequency, Active power to battery, Battery SoC (G2V, V2G)

Figure 2 depicts the battery scope view. The battery's graphical representation for the SoC, current, voltage, power utilization, energy utilization, as well as electrical charging cost in relation to time, is shown below. The characteristics parameter is shown in this simulation result after 10 s. The pulse signal identification unit is set to V2G inverter functionality when it is in discharging mode.

The electric power transfer processes through the battery to the electrical grid via power converters. The battery SOC, electrical power, and energy, as well as the time to

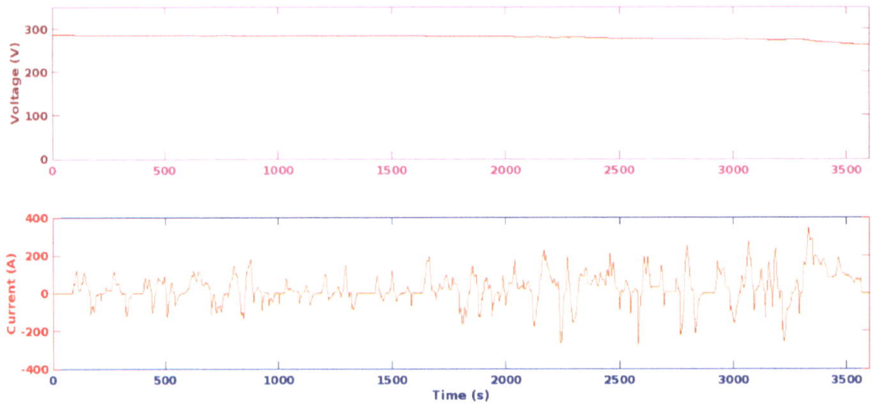

Fig. 3. Graphical Representation of Voltage and Current

park, are tracked. To avoid the synchronization problem, the electrical voltage waveforms obtained in the power Grid as well as the inverter have a similar magnitude field and phase difference. Figure 3 shows the waveform of voltage and current.

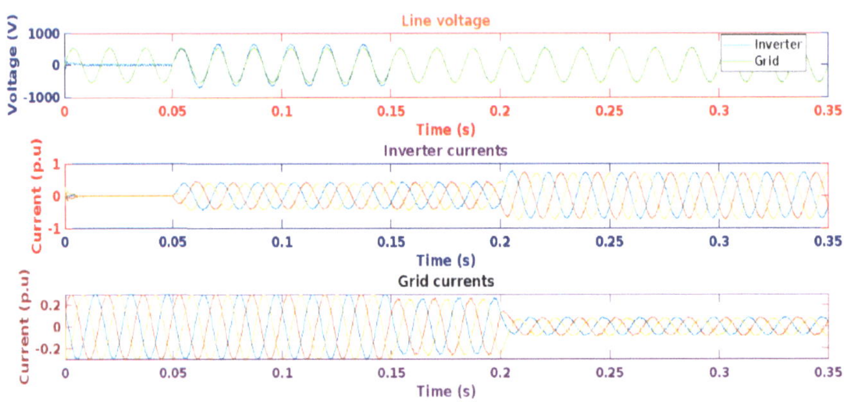

Fig. 4. Graphical Representation Grid voltage, current, and inverter current

The predicted period of time to finish the vehicle charging is determined by the battery SOC state. The electrical grid side inverter and the voltage side are displayed in Fig. 4. At 440 V and 50 Hz, the magnitude is the same. Both signals are in phase and synchronized.

4 Conclusion

The proposed concept of a V2G charging station demonstrates the ability to effectively manage the flow of power by considering the state of charge of the EV batteries that are linked to the station. This concept additionally enables the charging station to meet the energy requirements of electric vehicle consumers while simultaneously offering frequency management services to the power grid. The aforementioned characteristics are inherently linked to the significant IoT performance inside the V2G network of battery charging stations. The IoT plays a vital role in the auxiliary services sector, which is essential for ensuring the continued stabilization of power grids. V2G technology offers a distinct advantage by enabling the grid to achieve a more stable state in closer proximity to the demand endpoints. V2G technology facilitates cost reduction for end customers in their electricity use. In the lack of V2G technology, the process of energy acquisition necessitates a dependence on backup power plants. Due to this dependence, there is an escalation in peak electricity expenses as a result of the necessity to construct and uphold supplementary power facilities, incurring substantial financial burdens.

References

1. Wang, C., Liu, R., Tang, A.: Energy management strategy of hybrid energy storage system for electric vehicles based on genetic algorithm optimization and temperature effect. J. Energy Storage **51**, 104–314 (2022)
2. Safayatullah, M., Elrais, M.T., Ghosh, S., Rezaii, R., Batarseh, I.: A comprehensive review of power converter topologies and control methods for electric vehicle fast charging applications. IEEE Access **10**, 40753–40793 (2022)
3. Ronanki, D., Kelkar, A., Williamson, S.S.: Extreme fast charging technology—prospects to enhance sustainable electric transportation. Energies **12**(19), 1–10 (2019)
4. Mastoi, M.S., et al.: A study of charging-dispatch strategies and vehicle-to-grid technologies for electric vehicles in distribution networks. Energy Rep. **9**, 1777–1806 (2023)
5. Sarieddine, K., Sayed, M.A., Torabi, S., Atallah, R., Assi, C.: Investigating the security of EV charging mobile applications as an attack surface. ACM Trans. Cyber-Phys. Syst. **7**(4), 1–28 (2020)
6. García de Blanes Sebastián, M., Sarmiento Guede, J. R., Azuara Grande, A., Juárez-Varón, D.: Analysis of factors influencing attitude and intention to use electric vehicles for a sustainable future. J. Technol. Transfer **1**, 1–22 (2023)
7. Hopkins, E., Potoglou, D., Orford, S., Cipcigan, L.: Can the equitable rollout of electric vehicle charging infrastructure be achieved. Renew. Sustain. Energy Rev. **182**, 1–13 (2023)
8. Hossain, M.S., Kumar, L., Islam, M.M., Selvaraj, J.: A comprehensive review on the integration of electric vehicles for sustainable development. J. Adv. Transp. **2022**, 1–26 (2022)
9. Pardo-Bosch, F., Pujadas, P., Morton, C., Cervera, C.: Sustainable deployment of an electric vehicle public charging infrastructure network from a city business model perspective. Sustain. Cities Soc. **71**, 1–10 (2022)
10. Das, S., Acharjee, P., Bhattacharya, A.: Charging scheduling of electric vehicle incorporating grid-to-vehicle and vehicle-to-grid technology considering in smart grid. IEEE Trans. Ind. Appl. **57**(2), 1688–1702 (2020)
11. Hall, D., Lutsey, N.: Emerging best practices for electric vehicle charging infrastructure.In: The International Council on Clean Transportation (ICCT), Washington, USA, vol. 54, pp. 1–10 (2017)
12. Karolemeas, C., Tsigdinos, S., Tzouras, P.G., Nikitas, A., Bakogiannis, E.: Determining electric vehicle charging station location suitability: a qualitative study of Greek stakeholders employing thematic analysis and analytical hierarchy process. Sustainability **13**(4), 1–21 (2021)

Predictive Analysis of Telecom Customer Churn Using Machine Learning Techniques

K. Baby Lavanya[1](✉), M. D. Ismail Ansari[1], Venkateswarlu Gundu[2], G. Krishna Mohan[2], and B. Mouleswararao[2]

[1] Department of Computer Science and Applications, Koneru Lakshmaiah Education Foundation, Guntur, AP, India
lavanyakadiyala250@gmail.com

[2] Department of Computer Science and Engineering, Koneru Lakshmaiah Education Foundation, Guntur, AP, India

Abstract. Customer churn poses a significant challenge to the telecom industry as it directly impacts revenue and customer retention efforts. To tackle this concern, the present research suggests an all-encompassing machine learning oriented method for forecasting customer attrition within the telecommunications industry. The dataset is employed for training and testing the Machine Learning models, and their effectiveness is assessed through conventional metrics like accuracy, precision, recall, and the area under the curve of the receiver operating characteristic (AUC-ROC). Conducting an analysis of feature importance enables us to pinpoint the foremost factors that impact customer churn. The findings illustrate that the logistic regression model attains accuracy in its predictive capabilities for customer attrition. The approach offers a practical and accurate solution, aiding telecom companies in reducing churn, increasing customer loyalty, and ultimately improving business performance.

Keywords: Customer Churn · Machine Learning · Logistic Regression

1 Introduction

Churn, the phenomenon where customers switch to competing telecom services or terminate their subscriptions. In the telecom sector's rapid evolution, customer churn is a pressing issue causing revenue loss and relationship setbacks. With rising competition and expectations, predicting and curbing churn is now vital for telecom providers aiming to maintain market share and customer value. To tackle this, businesses are adopting advanced analytics and machine learning to uncover behavior insights and effective prevention statistics. Studies like Karvana [14] and V.Kavitha [15] use SVM, and Random Forests respectively for prediction. Prakash K.B.'s [18] presents an exhaustive manual towards data science and addressing practical aspects. Yajun Liu et al. [16] and others highlight ensemble learning, clustering, and models like Random Forest and XG Boost [17] for churn prediction. This study examines telecom customer churn using various

V. E. Balas et al. (Eds.): IEQIP 2023, LNNS 1029, pp. 100–106, 2024.
https://doi.org/10.1007/978-3-031-61929-8_15

ML models: SVM, XG Boost, KNN, Random Forest, Decision tree and Logistic Regression. Data preparation includes data cleaning, error correction, and feature engineering. We analyse correlations, split data(70% train, 30% test) and evaluate using accuracy, precision, recall, and F1–score. Logistic Regression standout, achieving 79% accuracy for churn prediction.

2 Literature Review

Various studies explore telecom churn prediction using ML models Paper. In [1], the author highlights telecom customer retention, comparing ML models: Gradient Boosting excelled, Random Forest and Logistic Regression were average, SVM underperformed, providing insights for retention strategies. [2] This study targets telecom churn management, using Bagging for efficient identification of potential churners, boosting profits and market share. Credibility is reinforced by author validation, adding weight to the research. [3] Emphasizing preprocessing and data balance, the author advocates an approach with SMOTE, feature extraction, and ensemble methods for accurate telecom churn prediction. Ada Boost excels Valuable insights for churn management. The author [4] examines practical applications of cognitive engineering in future computing. The author [5] recommends combining SVM with boosting for higher churn prediction accuracy, emphasizing continuous model development. Sai [6] finds better churn prediction with statistical tests and logistic regression, suggesting feature adjustments. Hemlata Jain [7] achieves High accuracy with Logistic Regression and Logit Boost, proposing a hybrid model. The study also plans to compare techniques like rough set, SVM, and neural networks on real datasets for insightful comparisons. M Panjasuchat [8] employs reinforcement learning for churn prediction, where XG Boost excels in accuracy, precision, recall, and F1. KNN demonstrates pattern stability, while DQN enhances predictions. Paper [10] examines random forest and deep neural networks, with a 32-layer network showcasing stable accuracy in identifying early customer outflow, aiding retention policies via marketing and services. The paper [9] sharing observations on total distance algorithm. In [11], Raja Gopal Kesiraju VLN highlights customer churn's significance and proposes using support vector machines for effective churn rate forecasting in complex telecom data. [12] by Praveen Lalwani showcases Adaboost and XGBoost as superior classifiers with future potential in reinforcement and deep learning. [13], Abhinav's study underscores Random Forest's accuracy in identifying at risk customers and churn factors, valuable for improving telecom services and understanding churn dynamics.

3 Machine Learning Models

3.1 Support Vector Machine

Support Vector Machines belong to a category of supervised machine learning algorithms utilized for tasks involving classification and regression. They are particularly effective in scenarios where the data in not only complex but also not linearly separate.

$$f(x) = sign\left(\sum_{i=1}^{n} \alpha_i y_i K(x, x_i) + b\right) \tag{1}$$

3.2 Logistic Regression

Logistic Regression is a fundamental and widely used statistical model in machine learning primarily used for binary classification tasks. Despite its name, Logistic Regression operates as a linear model that estimates the likelihood of an instance being a part of a specific class, relying on input features.

$$P(y = 1/x) = 1/(1 + e^{-z}) \tag{2}$$

3.3 XG Boost

XG Boost, short for Extreme Gradient Boosting, stands as a potent and extensively adopted machine learning technique known for its exceptional performance in both classification and regression assignment.

3.4 Random Forest

Random Forest combines many decision trees for strong predictions in tasks like categories or numbers. Each tree uses part of the data and random features, reducing errors. It's excellent for complex data and boosting accuracy.

4 Result Analysis

Analyzing telecom customer churn data from Airtel in Andhra Pradesh, India, using a 70–30 split for training and testing, yielded insights into customer behavior and factors influencing churn. The results obtained can guide Airtel's efforts in customer retention strategies, allowing them to proactively address issues that lead to churn. Several important features were considered to understand their impact on customer churn. The features included 'tenure' {T}, 'Online Security' {OS}, 'Tech Support' {TS}, 'Contract' {CT}, 'Paperless Billing' {PB}, and 'Monthly Charges'{MC}.

Based on the analyses conducted across various test cases, we can conclude that the combination of features 'tenure' (T), 'Online Security' (OS), 'Tech Support' (TS), 'Contract' (CT), and 'Monthly Charges' (MC) consistently demonstrated an accuracy equal to that of using all six features. In this particular case, where we considered {T-OS-TS-CT-MC}, we subjected it to a rigorous evaluation using a wide range of machine learning techniques, including decision tree, logistic regression, random forest, support vector machines (SVM), XG Boost and k-nearest neighbors (k-NN).

Upon evaluating the various machine learning models on the feature combination {T-OS-TS-CT-MC}, the logistic regression model emerged as the most accurate, yielding an accuracy of 0.7922. This indicates that the logistic regression algorithm was able to successfully predict customer churn with a high level of precision using the selected features.

In this research endeavor, Several machine learning techniques have been tested in the research for the prediction of customer churn as shown in the table (along with their accuracies). Based on the accuracies, observations can be made that Logistic Regression

Case-1: Model Evaluation using one feature

Feature	Accuracy
T	0.64
OS	0.60
TS	0.61
CT	0.63
PB	0.59
MC	0.62

Case-2: Model Evaluation using two features

Features	Accuracy	Features	Accuracy
T-OS	0.68	OS-MC	0.68
T-TS	0.65	TS-CT	0.68
T-CT	0.66	TS-PB	0.67
T-PB	0.62	TS-MC	0.66
T-MC	0.66	CT-PB	0.67
OS-TC	0.65	CT-MC	0.65
OS-CT	0.67	PB-MC	0.67
OS-PB	0.66	-----------	------------

Case-3: Model Evaluation using three features

Features	Accuracy	Features	Accuracy
T-OS-TS	0.69	OS-TS-PB	0.69
T-OS-CT	0.68	OS-TS-MC	0.72
T-OS-PB	0.70	TS-CT-PB	0.74
T-OS-MC	0.72	TS-CT-MC	0.71
OS-TS-CT	0.71	CT-PB-MC	0.70

(79.22%) and Support Vector Machine (79.08%) achieved the highest accuracies. However, remaining techniques like Decision Tree, Random Forest, XG Boost and KNN also recorded some better accuracies ranging from 73.21% to 76.57%.

Comparing proposed research with previous papers and presenting the findings in Table 1 and Table 2 is an effective way to showcase the uniqueness and contributions of telecom customer churn using machine learning techniques.

Case-4: Model Evaluation using four features

Features	Accuracy
T-OS-TS-CT	0.76
T-OS-TS-PB	0.75
T-OS-TS-MC	0.77
OS-TS-CT-PB	0.76
OS-TS-CT-MC	0.78
TS-CT-PB-MC	0.77

Case-5: Model Evaluation using five features

Features	Accuracy
T-OS-TS-CT-PB	0.76
T-OS-TS-CT-MC	0.79
T-TS-CT-PB-MC	0.77
T-OS-CT-PB-MC	0.78
T-OS-TS-PB-MC	0.78
OS-TS-CT-PB-MC	0.77

Case-6: Model Evaluation using six features

Features	Accuracy
T-OS-TS-CT-PB-MC	0.79

Table 1. Proposed ML Techniques

Machine Learning Technique	Accuracy
Decision Tree	0.7321
Random Forest	0.7628
SVC	0.7908
XGB	0.7733
KNN	0.7657
Logistic Regression	0.7922

Table 2. Comparison with Existing Methodologies

Paper Number	Machine Learning Technique	Accuracy
Paper [5]	Support Vector Machine	70%
Paper [8]	KNN classifiers and XG Boost	72%
Paper [12]	Ada boost and XG Boost	75%
Paper [13]	Random Forest	76%
Paper [14]	Support Vector Machine	68%
Paper [15]	Random Forest	72%
Proposed Model	Logistic Regression	79%

5 Conclusion

In conclusion, the strategies and methods mentioned above could be helpful for the future prediction and analysis of customer churn in the telecom industry. Changes can be made on model parameters and various features of customer behaviour for better results or accuracies. This progress helps for the customer-driven innovation and the business growth of the industry.

References

1. Gaur, A., Dubey, R.: Predicting customer churn prediction in telecom sector using various machine learning techniques. In: 2018 International Conference on Advanced Computation and Telecommunication (ICACAT), pp. 1–5. IEEE (2018)
2. Ali,M., Rehman, A.U., Hafeez, S., Ashraf, M.U.: Prediction of churning behavior of customers in telecomsector using supervised learningtechniques. In: 2018 (INC) on Computer, Control, Electrical, Electronics Engineering (2018)
3. Rizun, N., Baj-Rogowska, A.: Can web search queries predict prices change on the real estate market? IEEE Access **9**, 70095–70117 (2021)
4. Prakash, K.B.: Quantum meta-heuristics and applications Cognitive engineering for next generation computing: a practical analytical approach, pp. 265–297 (2021). https://doi.org/10.1002/9781119711308.ch10
5. Saran Kumar, A., Chandrakala, D.: A survey on customer churn prediction using machine learning techniques. Int. J. Comput. Appl. **975**, 8887 (2016)
6. Sai, B.N.K., Sasikala, T.: Predictive analysis and modeling of customer churn in telecom using machine learning technique. In: 2019 3rd International Conference on Trends in Electronics and Informatics (2019)
7. Jain, H., Khunteta, A., Srivastava, S.: Churn prediction in telecommunication using logistic regression and logit boost. Proc. Comput. Sci. **167**, 101–112 (2020)
8. Panjasuchat, M., Limpiyakorn, Y.: Applying reinforcement learning for customer churn prediction. J. Phys. Conf. Ser. **1619**(1), 012016 (2020)
9. Prakash, K.B.: Content extraction studies using total distance algorithm. In: Proceedings of the 2016 2nd International Conference on Applied and Theoretical Computing and Communication Technology,iCATccT 2016, art.no.7912085, pp. 673–679 (2017)

10. Kolomiiets, A., Mezentseva, O., Kolesnikova, K: Customer churn prediction in the software by subscription models it business using machine learning methods. In: CEUR Workshop Proceedings, vol. 3039, pp. 119–128 (2021)
11. RajaGopal Kesiraju, V.L.N., Deeplakshmi, P.: Dynamic churn prediction using machine learning algorithms-predict your customer through customer behavior. In: 2021 International Conference on Computer Communication and Informatics (ICCCI), pp.1–6. IEEE(2021)
12. Lalwani, P., Mishra, M.K., Chadha, J.S., Sethi, P.: Customer churn prediction system: a machine learning approach. Computing **104**(2), 271–294 (2022)
13. Thorat, A.S., Sonawane, V.R.: A random forest churn prediction model: an study of machine learning techniques for churn prediction and factor identif ication in the telecomsector. Scand. J. Inf. Syst. **35**(1), 818–824 (2023)
14. Karvana, K.G.M., Setiadi, Y., Syalim, A., Mursanto, P.: Customer churn analysis and prediction using data mining models in banking industry. In: 2019 internationalworkshop on big data and information security (IWBIS), pp.33–38. IEEE (2019)
15. Kavitha, V., Hemanth Kumar, G., Mohan Kumar, S.V., Harish, M.: Churn prediction of customer in telecom industry using machine learning algorithms. Int. J. Eng. Res. Technol. (IJERT) **9**(5), 181–184 (2020)
16. Liu, Y., Fan, J., Zhang, J., Yin, X., Song, Z.: Research on telecom customer churn prediction based on ensemble learning. J. Intell. Inf. Syst. **60**(3), 759–775 (2023)
17. Pallavi, H., Jadhav, G., Janagave, V., Kadam, S.: Customer churn prediction using machine learning algorithm
18. Prakash, K.B.: Data Science Handbook: A Practical Approach, pp. 1–453 (2022). https://doi.org/10.1002/9781119858010

Baker's Map Based Chaotic Image Encryption in Military Surveillance Systems

Nithya Paranthaman Rajendiran(✉)

Department of Networking and Communications, SRM Institute of Science and Technology, Kattankulathur, India
nithyar2@srmist.edu.in

Abstract. One of the primary problems in military surveillance systems is the safety of the visual data gathered by the sensor nodes installed in the different areas of the Visual sensor network and the remotely operated devices. There are several cryptographic algorithms available to secure the visual information. Several chaotic schemes worked well because they were more resistant to threats and had greater entropy levels. But, it takes a lot of time to finish the encryption step. Traditional approaches had the drawback that increasing encryption strength slowed calculation speed. This paper presents a scheme called "baker's map chaotic method based image encryption" to accelerate the encryption process in a network of visual sensors without weakening the encryption method. The findings indicate that the system's functionality is improved when the computation time is reduced to sixty seconds and the PSNR is increased to 15 dB. The performance of the chaotic method is compared by using key space analysis, key sensitivity analysis, correlation coefficient analysis, noise analysis, Entropy analysis, and Speed analysis. The IBM SPSS statistical tool was also used to do statistical analysis.

Keywords: Military Surveillance · Chaotic Encryption · NPCR · UACI · PSNR

1 Introduction

The primary aim of encryption based on chaos is to use different dynamic structures to produce a string of numbers from which a single value can be chosen randomly. In the case of encryption using dynamic systems, recovering the original data fast without the original key is not possible. However, it can be restored to its original shape by setting the proper initial conditions. The invader cannot use knowledge of the original RGB image to deduce the precise initial conditions necessary to circumvent it. There is a direct correlation between the chaotic map dimension and the complexity of chaotic procedures. The image was encrypted using dynamic structures of this type. Several different 2D chaotic maps can encrypt images. The design parameters determine the degree of randomness and the security of the key creation. Taking elements from Baker's algorithm, the revolutionary hybrid scattering chaotic method provides a new and exciting approach to chaotic processes in visual sensor network represented in Fig. 1.

V. E. Balas et al. (Eds.): IEQIP 2023, LNNS 1029, pp. 107–115, 2024.
https://doi.org/10.1007/978-3-031-61929-8_16

Fig. 1. Visual sensor network of the intelligent military surveillance system, with sensors located in key locations

2 Related Work

Many image encryption algorithms have been proposed recently. Some algorithms have larger keys and more security. Fridrich et al. (1998) proposed a image encryption based on chaotic technique that consists of two different procedures: confusion of pixels and diffusion. The earlier algorithm used a 2D chaotic model to permute the pixels of a simple image, whereas the latter sequentially adjusts the values (grey level) of each pixel. This framework played a significant role in developing numerous chaos-based image ciphers. Ansari et al. (2008) suggested a novel method for encrypting an image utilizing chaotic maps in the Frequency Domain. A 2D baker's map is used to shuffle the image after it has been analyzed using the Discrete Cosine Transform (DCT). Two baker's maps are employed, the first utilizing the primary set of keys and the second applying a Gaussian image created with mean and variance. Both the gain of baker's maps and DCT are repeatedly XORed with one another. A random sequence based on Gaussian distribution is generated by a number generator to create the scattering pattern. The proposed encryption technique employs two baker's maps and can handle key spaces of up to 128 bits. Yaobin Mao et al. (2004) presented a simple image encryption system based on Baker maps, which involved expanding the Baker map of two dimension into three dimensions with a more chaotic structure. In addition, a significant acceleration of the encryption procedure was accomplished. Chen et al. (2004) introduced a 3D cat map method to scramble the placements of pixels in the image. An additional chaotic map method was applied to baffle the relation between the cipher and the plain image. Thus, the defense against analytical and quantitative threats is drastically increased. Lian et al. (2005) demonstrated several weak cipher keys utilizing the cat and baker models. In addition, the key space of these models is less than that of the regular model. Consequently, they proposed employing a standard model for confusion while maintaining the logistic model for image pixel diffusion.

3 Baker's Map Chaotic Method Based Image Encryption

The plain-text image is subjected to a randomization method that performs the flattening and stretching operations. The Number of iterations directly correlates to the image's unpredictability and security, and increasing the number of iterations improves security but lengthens the time required to decrypt a message. Figure 2 is a diagrammatic illustration of the chaotic approach based on the image encryption of General Baker's map. Figure 3 shows a diagrammatic illustration of (A) General Baker's map chaotic technique-based image encryption. The Baker map randomly scatters the image in all directions, initially in a horizontal direction until it reaches the end of the block and then in a vertical order until it reaches the row directly above the currently visible one. Below is the baker function, where a and b represent the system parameter values a = 1.4 and b = 0.3, and S_{n1} and S_{n2} define state parameters in Fig. 3(b). There are two possible chaotic sequences, S_{n1} and S_{n2}, because the system is chaotic. Now, the image is encrypted using these key values generated using the Eq. 1 and 2.

$$S_{n1+1} = 1 - S_{n1^2} + S_{n2} \tag{1}$$

$$S_{n2+1} = bS_{n2} \tag{2}$$

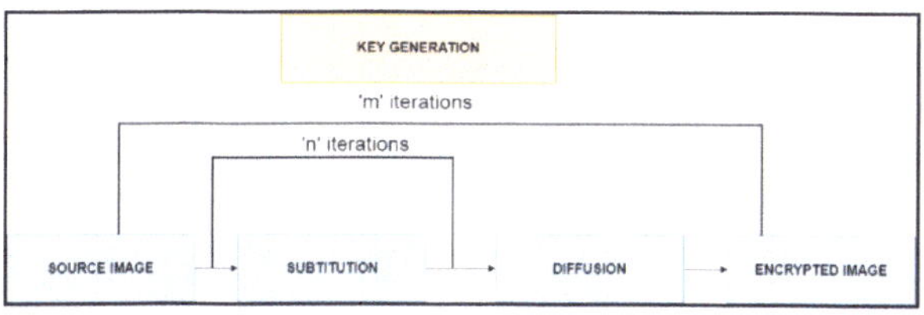

Fig. 2. Schematic representation of General Baker's map chaotic method based image encryption

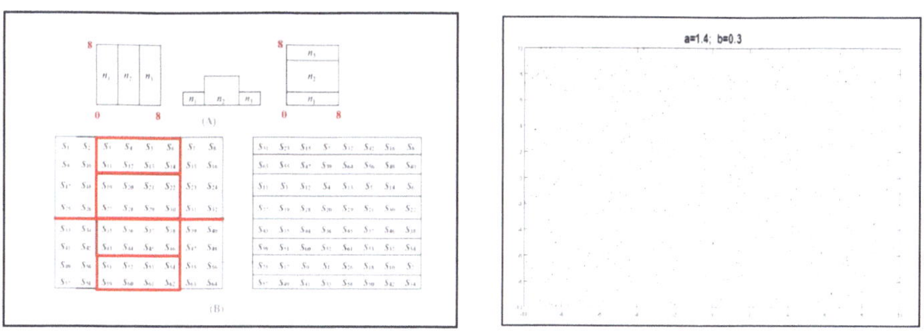

Fig. 3. Schematic representation of (A) General Baker's map chaotic method-based image encryption (B) Discretized Baker's map chaotic method-based image encryption. (C) Baker's Chaotic Map

It can be seen that the values of S_{n1} and S_{n2} obtained before the key generation in the x-direction are used. However, key generation in the y-axis is independent of S_{n1}

and uses the current and previous S_{n2} values. When setting the default values for the control parameters, it is essential to ensure they aren't so far off that the iteration over the fixed point stops at 0. As determined by random testing, the values a = 1.4 and b = 0.3 produced superior results without eliminating the iterations. S_{n1}, b, and S_{n2}were found to take on initial values between 0 and 1. Figure 4 depicts the chaotic distribution that resulted from using Baker's map. Baker's encryption technique is used to encrypt the live samples, and the results are shown in Fig. 4.

Fig. 4. Baker chaotic approach for real-time Visual sensor network images

4 Performance Evaluation

Numerous experiments have been carried out to determine the efficiency of Baker's map chaotic method-based image encryption. Figure 4 illustrates the application of a baker's map chaotic technique-based image encryption for a visual sensor network.

4.1 Key Sensitivity Analysis

More sensitive cryptography is needed to detect private key modifications. When the decryption key is significantly changed, the outcome is virtually equal to the cypher image but not the original. Key sensitivity is assessed by: the Unified Mean of All Pixel Level Change (UACI). Changing pixels or key values affects encryption security. Double-encrypting an image increases key sensitivity. First, encrypt C1 using the chaotic approach. It's encrypted again with C2 keys. First, image encryption adds sharpness. Second-round encryption modifies images' least effective pixels. Figure 5a represents the Graphical bar chart representation of Baker's map chaotic method based image encryption NPCR, UACI values and Table 1 depicts the NPCR and UACI for Baker's map chaotic method-based image encryption.

4.2 The Coefficients of Correlation

The correlation between neighboring pixels determines the encryption method's reliability. For the encryption to work, the relationship between adjacent pixels must not be linear, as shown by correlation coefficients that tend toward zero. Calculate correlation coefficients by fixing N in each direction. The unencrypted image has higher correlation coefficients in all directions. Table 1 presents the Cross Relation (R Band, G Band, and B Band) for the 512 × 512 Sample Image Encrypted Using the Baker's Map Chaotic Method. Figure 6 shows the directional cross-correlation matrix based on image encryption for the baker's map chaotic approach. First, calculate neighboring pixel's covariance and variance to calculate cross-correlation given in Eq. 3.

$$corellation\ coefficient(i, j) = \frac{covariance}{\sqrt{variance(i)}\sqrt{variance(j)}} \tag{3}$$

where i and j denote two neighboring pixel's black and gray levels.

4.3 Peak Signal to Noise Ratio (PSNR)

The Peak Signal to Noise Ratio (PSNR) is a statistic used to evaluate the quality of the encrypted image on the receiving end of the transmission. The Eq. 6 gives the values of PSNR.

$$Peak\ Signal\ to\ Noise\ Ration = \frac{10 \log M*N(255)^2}{\sum_{x=0,y=0} (function_1(x, y) - function_2(x, y))^2} \tag{4}$$

Table 1 displays the PSNR values for the R Band, G Band, and B Band encrypted image with Baker's Map Chaotic Method. Figure 5b depicts the Graphical bar chart representation of Baker's map chaotic method-based image encryption PSNR (R Band, G Band, and B Band) for the encrypted image.

4.4 Entropy Information

The entropy of an image is the degree of uncertainty in the information deduced from its mean pixel value. Theoretically, an image's pixels with values between 0 and 255 should have an entropy value of eight. It can be calculated using the following Eq. 5,

$$Entropy(x) = \sum\nolimits_{a=1}^{n} probability(x_i)\log_2 probability(x_i) \tag{5}$$

The standard value of entropy is 8 and here the values of Entropy obtained was nearer to 8.

4.5 Analysis of Computation Time

Furthermore, when encrypting images in real-time, other parameters appear more significant. Speed analysis is crucial for multimedia image encryption to take place in real-time. All tests used seventh-generation Intel Core i5 processors with eight gigabytes of RAM and a half-terabyte of hard disc space. The computation time analysis of Baker's map chaotic encryption method is presented in Table 2. The computation time was found to be around 60 s.

Table 1. Analysis of NPCR and UACI, Cross Correlation and PSNR of the Baker's map chaotic method based image encryption for the Sample Images of size 512 × 512

Sample Image 512 × 512	NPCR (%)	UACI (%)	Baker's map chaotic method based image encryption				PSNR of G Band[dB]	PSNR of B Band
			Correlation in R Band	Correlation in G Band	Correlation in B Band	PSNR of R Band[dB]		
1	99.60	29.94	4.35×10^{-4}	–0.0035	6.45×10^{-5}	14.78	14.22	13.19
2	99.59	31.88	–0.0027	–0.0024	–0.0026	12.53	11.81	7.97
3	99.60	32.19	–0.0035	–0.0015	–0.0037	12.26	11.44	10.87
4	99.62	31.83	–0.0021	–0.0028	–0.0024	12.31	11.57	9.027
5	99.61	35.17	0.0029	0.0016	-8.98×10^{-4}	10.19	10.20	9.60

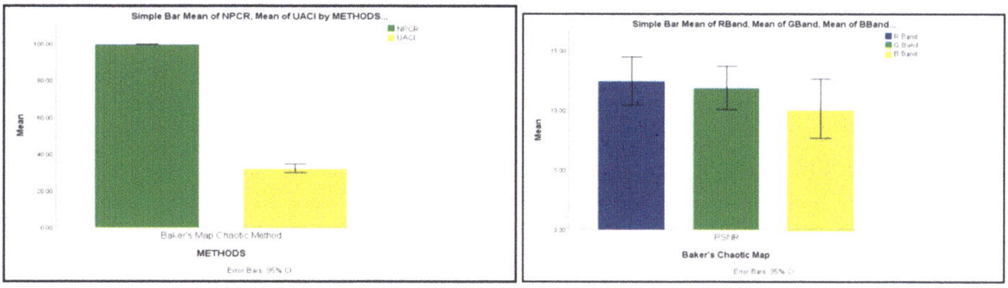

(a) (b)

Fig. 5. Comparison of Baker's map chaotic method based image encryption with the Sample image of size 512 × 512 (a) Graphical bar chart representation of NPCR, UACI (b) PSNR (R Band, G Band, and B Band)

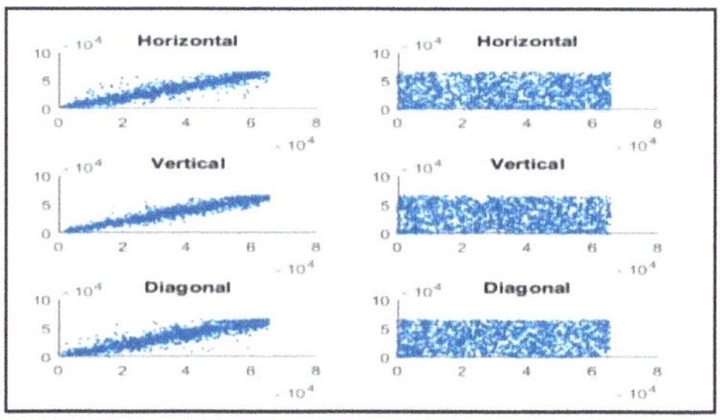

Fig. 6. Directional cross-correlation matrix for the Baker's map chaotic method based image encryption

Table 2. Entropy analysis and Computation Time on both the original and the encrypted images with the baker's map chaotic technique.

Sample image of size 512 × 512	Entropy of the Original Image	Entropy of Baker's Method	COMPUTATION TIME(S)
1	7.6565	7.9997	62.91
2	7.7247	7.9997	59.03
3	7.7050	7.9997	56.75
4	7.6015	7.9997	58.52
5	7.4677	7.9997	61.27

5 Conclusion

The typical framework of Baker's map based chaotic image encryption technique has been investigated. To create a more random appearance for the pixels, the original Chaotic Baker map is used. The proposed framework demonstrates less impressive performance analysis for cross-correlation, entropy, peak signal-to-noise ratio, and computational time analysis. The main objective of this inquiry is to find an innovative method that can significantly reduce the amount of time required to encrypt data. The computation time was found to be around 60 s. Future study must be conducted on the proposed methodologies using a variety of real-time images received from various other surveillance-based systems. The proposed adaptive security can be implemented in vision sensors depending on the power budget that is available.

References

Nithya, R., Dhanasekaran, D.: Novel dominant color subband image encryption in visual sensor network for smart military surveillance system. Traitement du Signal, **39**(3) (2022)

Paranthaman, R.N., Dhanasekaran, D.: Implantable electronics: real-time adaptive image security of smart visual sensor nod. In Reinvention of Health Applications with IoT, pp. 141–154. CRC Press

Nodal Level Adaptive Security for IOT-Enabled Sensors through Dynamic Reconfigurable Encryption R. N. Paranthaman, D. Dhanasekaran Published 2019 Computer Science International Journal of Engineering and Advanced Technology

Subramanyan, B., Chhabria, V.M., Babu, T.S.: Image encryption based on AES key expansion. In: 2011 Second International Conference on Emerging Applications of Information Technology, Kolkata, pp. 217–220 (2011).https://doi.org/10.1109/EAIT.2011.60

Radhadevi, P., Kalpana, P.: Secure image encryption using AES. Int. J. Res. Eng. Technol. **1**(2), 115–117 (2012)

Akram, B., Oussama, B., Houcemeddine, H., Safya, B.: Selective image encryption using DCT with AES cipher. In: Proceedings of the NETCOM, pp. 69–74 (2014). https://doi.org/10.5121/csit.2014.41306

Chen, T., Zhang, M., Wu, J., Yuen, C., Tong, Y.: Image encryption and compression based on kronecker compressed sensing and elementary cellular automata scrambling. Opt. Laser Technol. **84**, 118–133 (2016). https://doi.org/10.1016/j.optlastec.2016.05.012

Fan, H., Zhang, C., Lu, H., Li, M., Liu, Y.: Cryptanalysis of a new chaotic image encryption technique based on multiple discrete dynamical maps. Entropy **23**(12), 1581 (2021). https://doi.org/10.3390/e23121581

Mokhnache, A., Ziet, L.: Cryptanalysis of a pixel permutation based image encryption technique using chaotic map. Traitement du Sig. **37**(1): 95–100 (2020). https://doi.org/10.18280/ts.370112

Hazarika, N., Borah, S, Saikia, M.: A wavelet based partial image encryption using chaotic logistic map. In: 2014 IEEE International Conference on Advanced Communications, Control and Computing Technologies Ramanathapuram, India, pp. 1471–1475 (2014).https://doi.org/10.1109/ICACCCT.2014.7019347

Xiang, T., Wong, K.W., Liao, X.: Selective image encryption using a spatiotemporal chaotic system. Chaos: an Interdisc. J. Nonlinear Sci. **17**(2), 023115 (2007). https://doi.org/10.1063/1.2728112

Kuppusamy, K., Thamodaran, K. (2012). Optimized partial image encryption scheme using PSO. In International Conference on Pattern Recognition, Informatics and Medical Engineering (PRIME-2012), Salem, India, pp. 236- 241. https://doi.org/10.1109/ICPRIME.2012.6208350

Goel, A., Chaudhari, K.: Median based pixel selection for partial image encryption. In: 2016 Sixth International Conference on Image Processing Theory, Tools and Applications (IPTA), Oulu, Finland, pp. 1–5. IEEE (2016). https://doi.org/10.1109/IPTA.2016.7820931

Li, J., et al.: A partial encryption algorithm for medical images based on quick response code and reversible data hiding technology. BMC Med. Inform. Decis. Mak. **20**(14), 1–16 (2020). https://doi.org/10.1186/s12911-020-01328-2

Parameshachari, B.D.: Logistic sine map (LSM) based partial image encryption. In: 2021 National Computing Colleges Conference (NCCC), Taif, Saudi Arabia, pp. 1–6 (2021).https://doi.org/10.1109/NCCC49330.2021.9428854

Pandurangi Ramacharya, B., Patil, M.R., Keralkar, S.: Fast partial image encryption with fuzzy logic 959 and chaotic mapping. Evol. Intel. (2022). https://doi.org/10.1007/s12065-021-00693-

Saravanan, S., Sivabalakrishnan, M.: An image encryption scheme using chaotic baker map. J. Comput. Theor. Nanosci. **17**(5), 2130–2135 (2020). https://doi.org/10.1166/jctn.2020.8859

Cyber Security Investigation of GPS-Spoofing Attack in Military UAV Networks

N. Vanitha[1(✉)] and S. K. Rakshun Selvaa[2]

[1] Department of AI and ML, Coimbatore Institute of Technology, Coimbatore, Tamil Nadu, India
vanitha@cit.edu.in

[2] Department of Information Technology, Dr. N.G.P. Arts and Science College, Coimbatore, Tamil Nadu, India

Abstract. Unmanned Aerial Vehicles (UAV) in military are prone to attacks by hackers and terrorists and this concern deals with a Nation's security. GPS-Spoofing is a technique used by attackers to bring down a UAV to ground or to capture drone and perform Reverse Engineering which posses a key threat to Defense Systems and software suites being stolen. An UAV is captured when a spoofer gains the ability to eventually specify the UAV's position and velocity estimates. Both overt and covert spoofing strategies are estimated. GPS-Spoofing incidents have led to Military drones being captured in overseas territories while flying in restricted airspace. GPS-Spoofing is a threat to UAV's as they misguide the drones to think enemy bases as home base and they fall as a prey to the wrong hands. The paper aims to Explore through the theoretical frameworks to counter GPS-Spoofing with enhanced security and reliable navigation techniques. Averting a GPS-Spoofing attack through the integration of inertial as well as sensor reference data is theorized in this paper further.

Keywords: UAV · GPS-Spoofing · Reverse Engineering · Military

1 Introduction

UAV or Unmanned Aerial Vehicles are aircraft without onboard pilots, crew and civilians. UAV posses a greater advantage in logistics of various fields from Military usage to medical emergencies. Drone technology combined with artificial intelligence and machine learning can outperform human logical analysis in Military-Grade operations. Drones can be deployed in fields are there is a high risk probability for human lethally and fields where the human's reflexes are slow enough that modern sophisticated drone technology can take over the complex analysis of various maneuverable tasks.

UAV's posses greater advantage over payload delivery autonomously and over surveillance capabilities. UAV are being used in agriculture to water the crops, being tested in food and goods delivery capabilities, transport bulk payload through international airspaces, aerial photography and even for entertainment purposes. Drones plays a major role in Military applications which is discussed further below the paper [1].

V. E. Balas et al. (Eds.): IEQIP 2023, LNNS 1029, pp. 116–124, 2024.
https://doi.org/10.1007/978-3-031-61929-8_17

1.1 Application of Drone Networks in Military

Drones plays a major role in military perspective during combat missions. Drones are used in air force and navy to undergo reconnaissance missions and stealth operations. Drones are being deployed by countries that are advanced in military operations. Drones can reduce Human error associated with target manipulating and striking probability on enemy targets. Drones can carry payload like missiles, Anti-Tank Guided Missiles, bombs and etc. They can be also used in patrol, search and rescue missions, Military logistics where Human precision is being overtaken by Machine Learning and Artificial Intelligence infused UAV's. Border protection using military UAV's are far efficient and way faster and precise. Drones take a lion's share in their destruction power equally combatable towards to 4th Gen Fighter jets. They can cruise over thousands of miles towards the target to be screened. Drones don't require a Human Pilot as they are capable to hit Targets once programmed on ground. Information Technology plays a major role in designing Software for Drones and even during Designing of the drones. Military Drones are capable of High- altitude Flights, Evasive Maneuvers, Target manipulation, Weapons Deployment and their Flight Control Computer primarily needs the help of Information sciences to perform the above mentioned tasks. Drones are highly dangerous and plays as a potential threat to opponents when the threat is in their side. They can be deployed in anti-terror operations and covert operations. They posses' faster turnaround times which makes them suitable for reconnaissance missions. Drones can be engaged in Advanced Surveillance in Par with humans due to the ability to rapidly Deploy them. One of the most important role of military drones is counter insurgency by drones collecting the data from the combat area. They can provide good tactical 360-degree coverage about actionable intelligence which could be used by military to confirm the complexity of the threat possessed by the opponent. The future 6th or 7th Gen fighter jets have the capability to be classified as Drones. Drones can be expandable to be worked as a team with present Gen fighter jets to analyze data about the target to be attacked and the data can be distributed to the fighter jets for further attacks. Drones plays a major role to be compatible with Human Intelligence. There is no surprise military UAV has the potential to replace humans in the fighter cockpit deck with their intelligent technological advances [2].

1.2 Significance of Military Drones

Defense logistics is more complicated as they need to monitor and assess the enemy movements continuously. Drones being deployed in military can reduce the complexity of military network systems as more automation can also offer higher degree of protection against the risk of losing a human life. Military UAV's can work in collaboration with humans by fetching data that are nearly impossible for a human to collect in personal as they posses high risk. One of the primary advantages of drones is their size. UAV's don't require size standards and even a drone the size of a fighter jet or drone the size of a human hand are still capable of fetching enemy data intelligence. Technology is a key factor in critically managing UAV's which are remotely operated either more than a kilometer or hundreds of kilometer far from the ground-base. The precision of drones are far more better than humans with present developing technologies like Real-Time Kinematic by which the precision level were to be in centimeter-level. The probability of target being missed in a combat mission is a very low possibility. Drones strike mechanism can also be trained via artificial intelligence which doesn't need hefty training phases that humans endure. Drones can carry weapons in their bomb-bay as a payload and can fly until the strike zone and can return safely due to their higher analyzing and computational systems embedded in their processors. Military UAV's get their major role in replacing future Fighter jets into autonomous UAV's with little to no pilot or ground crew help as the future technology will be capable of fully autonomous drones.

1.3 Objectives

The primary objective of the paper is to perform security investigation of GPS Spoofing attacks on UAV communication.

2 Background Study

Various literatures from national and international authors are mentioned in the Table 1.

Table 1. Background Study

S.No	Authors	Year & Journal	Techniques	Observation
1	Hosam Alamleh, Nicholas Roy	2021 IEEE International IOT, Electronics and Mechatronics Conference (IEMTRONICS)	Breaking wi-fi systems to hack drones	Successful counterfeit signals were sent as input
2	Sait Murat Giray	2013 6th International Conference on Recent Advances in Space Technologies (RAST)	Signal spoofing	Attempt to capture drone was done with interfering signals
3	Wenxin Chen, Yingfei Dong, Zhenhai Duan	2022 IEEE 19th Annual Consumer Communications& Networking Conference(CCNC)	Redirecting a drone	Autopilot was over-ruled
4	João Gaspar, Renato Ferreira, Pedro Sebastião, Nuno Souto	2018 Global Wireless Summit(GWS)	GPS-spoofing	Drones flight plan was successfully altered

These above mentioned techniques were used by researchers to attack a drone and gain access to either the co-ordinates of the drone or to control the drone and to bring them to the ground. Military drone's posses the highest threat of being maliciously hacked or hijacked due to their high destruction power and precise intelligence gathering ability.

2.1 Attacks on Military Drones

Even the with most sophiscated technology drones are still prone to hijacking. A good example of a Military drone being hijacked is the Iran-US RQ-170 Sentinel incident.

This capture occurred on 5 Dec ember 2011 when a drone based on united states entered Iranian airspace which lead a threat to Iran. So the Cyberwarfare unit of Iran had brought down the drone made by American defense contractor Lockheed Martin and reverse engineered it to make their own version of autonomous UAV with the capabilities far more than the captured drone. This incident was helpful with GPS-Spoofing where the drone was successfully able to be manipulated to be deceive that the Iranian airspace to be US airspace and the eventually made the drone to land with minimal damage in Iran made thinking the drone that it had been landed in American home-base. This capture was eventually a significant leap in bringing down Military drones but they possess a greater risk of technology and software suites being stolen [3]. Figure 1 shows the RQ-170 Drone.

Fig. 1. RQ-170 Drone. Image source via: LOCKHEED MARTIN website

This drone is the captured RQ-170 sentinel manufactured by Lockheed Martin and it was reverse engineered by Iranian Militaryfirm. Figure 2 shows The reverse engineered version of the captured drone by iran.

Fig. 2. Image source via: Tasnim news

3 Introduction to GPS and GPS-Spoofing

Global positioning system is a constellation of satellites which acts as a radionavigation system. Four or multiple GPS satellitesprovides geolocation and information about time precisely to a GPS receiver. The malicious acts come into play as the GPS is freely accessible and mostly unencrypted to any GPS receiver. One such potential threat to GPS based systems are GPS- spoofing. GPS-spoofing is a unauthorized form of intervening systems that rely on GPS to deceive it is in false location by broadcasting counterfeiting GPS signals. It's attack ranges from drones to even larger sea vessels. GPS spoofing attacks are even a potential threat to power grids by intervening the Cyber-physical systems in Time synchronization. In june 2017, 20 ships in Black sea witnessed GPS based anomalies that the vessel are to be transpositioned miles from their true location. Countries such as Norway made spoofing during NATO forces exercising drills in oceans which had lead the ship to be collided. GPS spoofing plays a major threat to the safety and viability of the GPS based systems [4, 7, 8].

3.1 Overt Capture

Overt capture is a form of spoofing approach where there is no attempt for the attack to be concealed. The counterfeit signals are broadcasted with higher intensity than the signal coming from the GPS satellites with a spoofer device. The authentic signals from satellites are made to be lost and the intervening signals cause the drone to rely on the spoofing signals as their primary GPS input source signals. The problem faced by this approach is that it can be defeated by performing simple checks by the flight computer in drones. The majority of receivers that current time drone use are not bothered about performing anti-spoofing since spoofing is least a threat probability in drone flight plan but a specified threat to a certain targeted drone can cause the loss of the drone and even technological loss by exploiting the captured drones by malicious hackers or hijackers.

3.2 Covert Capture

The covert captured is the most technically advanced capturing of drones where the drones are equipped with systems that detect the oncoming spoofing attempts. The primary objective of covert capture is to not trigger the spoofing detection mechanism.

Navigation systems of advanced that are used for military operations are embedded with multiple redundant technologies to atmost avoid the GPS spoofing attack. To be considered as covert attack on navigation systems, some of the conditions to be satisfied are to evade alarming the GPS spoofing detectors onboard UAV, jamming-to-noise analysing within GPS based receivers, monitoring frequency unlock under the GPS based systems and testing navigation system's state estimator. Covert capture is done via counterfeiting signals to navigation systems without undetectable gradually. Post covert capture a drone can be made to deliberately crash onto obstacles or bring down to exploit the resources of drones and even made to reverse the original mission to even counter attack against the primary objective of any Military combat missions.

4 Security Investigation of UAV Networks

The above mentioned methods are therefore most crucial in the rule of thumb for destruction of UAV and reverse engineer the non- patentable military tactics, technologies, software suites, hardware suites, communication systems, new embedded onboard innovative yet competitive advantaged systems. Military drones being GPS-spoofed and manipulated could posses a deadly threat to the original manufacturer and operator as they can be turned into better competitor to the original origin of the nation by acquiring the drone technology that the captured drones posses. Drones but still have many possibilities to be implemented to avert an attack by malicious hijackers. They can be implemented either in Hardware or Software with the help of information technology. Hardware based implementation of drones are more redundant and safer as they are immune to interference and software can made to be redundant as hardware with special techniques.

4.1 The Inertial Navigation System for Attack Avoidance

The primary co-ordinate measuring system used by submarines is the inertial navigation system where the onboard computers or even mechanical devices are used to keep track of the inertial data using gyroscopes from the fixed stationary point. The inertial guidance system are accurate upto less than 150 h of mission hours which is a far greater time to avoid any potential threat possessed by changing the GPS-co-ordinates. When an anomaly is found my drone flight computers by sensing a sudden change in co-ordinates, the inertial navigation systems can over-rule GPS systems until reaching a home-base and can be safely recovered by radar assisted landing. Multiple hardware systems can therefore increase the redundancy of the drones avoidance towards potential GPS-based attacks. This technique can be aided with the help of Geomagnetic and therefore the new technique is called Geomagnetic aided Inertial Navigation System (GAINS). This technology makes use of Geomagnetic Information and it makes way for the future Gen passive navigation system. With these hardware redundant backup systems GPS-Spoofing can be averted by military and commercial UAV's [5].

4.2 The Astro-Inertial Reference Systems for Attack Avoidance

One of the technologies used in B-2 and sr-71 blackbird the fastest supersonic flight in history used in Military and by CIA of America is astro-inertial navigation system.

This system used in planes keep track of 52 stars in various constellations seen around the globe in memory of the astro-inertial reference system. This system can be helpful when it can over-rule GPS-spoofing and use stars to calculate it 's position and make an alternate flight plan to escape the GPS –anomaly causing airspace. Astro-inertial systems use special sensors fitted with telescopes that can pinpoint approximately the airspace where the drone is currently flying. It's Accuracy is estimated to be 90 m and the usage of astro-inertial navigation for course guidance as a backup to GPS can increase the Drone's safety level and also decreases the chances of being spoofed maliciously [6].

4.3 Inertial Reference System Using Software Expertise

The flight computer of drones can be used to keep track of flight plan by two redundant systems. The first system keeps track of the co-ordinates based on the input to the controls such as speed, altitude and trajectory of the drones and second system keeps track of the GPS co-ordinates. When there is a GPS spoofing attack the second system gets spoofed as it detects the spoofed GPS-co- ordinates. The flight computer now gets alerted due to sudden alteration in the second system and there is a mismatch with the calculated flight co-ordinates in the first system. The flight computer analyzes the two systems if there is a mismatch in the systems the drone gets its flight plan according to first system which keeps track of previous path and travels according with reference to the previous flight path. This method can be helpful to drones for short–medium range missions and can be saved from hijackers and hackers and by using the system-1 data the drone gets retraces to its original path and its gets away from GPS-spoofing threat. Figure 3 shows the implementation of avoiding an attack.

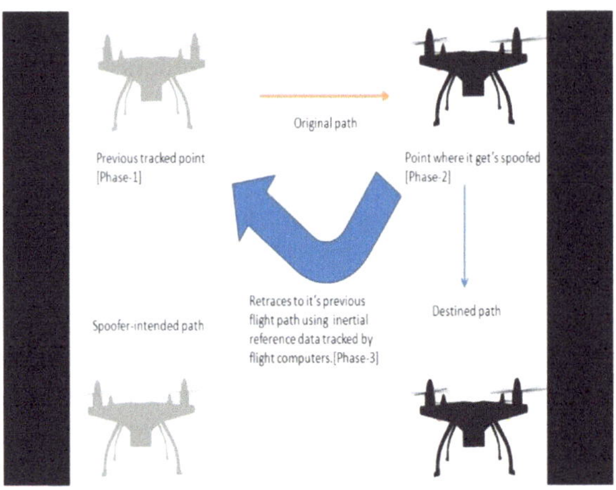

Fig. 3. Attack avert procedure using Inertial information tracking

4.4 Security Investigation of UAV'S Against Security Principles

CIA which stands for confidentiality, integrity, availability forms the important factors within security of the information and communication technology. Various examples of attacks are analyzed with this framework of CIA. Table 2 reveals that GPS- spoofing and GPS Jammer attacks are the most vulnerable attacks that compromises the most of the Security principles.

Table 2. CIA Analysis

Attacks	Confidentiality	Integrity	Availability
GPS-spoofing	✔	✔	✗
WI-FI spoofing	✔	✗	✗
DOS	✔	✗	✗
Man-in-the-Middle Attack	✗	✗	✗
GPS Jammer	✔	✔	✗
Malware Based Cyber-Attacks	✗	✗	✗

5 Conclusion and Future Work

This paper investigates the significance of Military Drones and the threat faced by the drone by hackers or hijackers. Military drones as well as commercial drones are facing hundreds of threat and malicious attacks. By neglecting the need for an anti-Spoofing system can lead to loss of drones and technology of the drone. GPS-spoofing avoidance system must be enhanced in every medium- long range UAV's which can save them from hackers and hijackers. In future Machine-learning algorithms can be used to identify the GPS-spoofing attacks from the data sets of drone networks.

References

1. Jinke, H., Guoqiang, F., Boxin, Z.: Research on autonomy of UAV system. In: 2021 IEEE 4th International Conference on Automation, Electronics and Electrical Engineering (AUTEEE), pp. 745–750 (2021). https://doi.org/10.1109/AUTEEE52864.2021.9668730
2. Schneiderman, R.: Unmanned drones are flying high in the military/aerospace sector [special reports]. IEEE Signal Process. Mag. **29**(1), 8–11 (2012). https://doi.org/10.1109/MSP.2011.943127
3. Restituyo, R., Hayajneh, T.: Vulnerabilities and attacks analysis for military and commercial IoT drones. In: 2018 9th IEEE Annual Ubiquitous Computing, Electronics & Mobile Communication Conference (UEMCON), pp. 26–32 (2018). https://doi.org/10.1109/UEMCON.2018.8796596
4. Renyu, Z., Kiat, S.C., Kai, W., Heng, Z.: Spoofing attack of drone. In: 2018 IEEE 4th International Conference on Computer and Communications (ICCC), pp. 1239–1246 (2018). https://doi.org/10.1109/CompComm.2018.8780865

5. Feng, H., Yang, Z., Fang, J.: Simulation design of geomagnetic aided inertial navigation system. In: 2008 2nd International Symposium on Systems and Control in Aerospace and Astronautics, pp. 1–5 (2008).https://doi.org/10.1109/ISSCAA.2008.4776399
6. Eric Long, National Air and Space Museum, Smithsonian Institution," Nortronics NAS-14V2 Astroinertial Navigation System" (2012)
7. Vanitha, N., Padmavathi, G.: A study on various cyber-attacks and their classification in UAV assisted vehicular Ad-Hoc networks. In: Ganapathi, G., Subramaniam, A., Graña, M., Balusamy, S., Natarajan, R., Ramanathan, P. (eds.) Computational Intelligence, Cyber Security and Computational Models. Models and Techniques for Intelligent Systems and Automation. ICC3 2017. Communications in Computer and Information Science, vol. 844, pp. 124–131. Springer, Singapore (2018). https://doi.org/10.1007/978-981-13-0716-4_11
8. Vanitha, N., Padmavathi, G.: A comparative study on communication architecture of unmanned aerial vehicles and security analysis of false data dissemination attacks. In: 2018 International Conference on Current Trends towards Converging Technologies (ICCTCT), Coimbatore, India, pp. 1–8 (2018). https://doi.org/10.1109/ICCTCT.2018.8550873

IoT Based Enhanced Safety Monitoring System for Underground Coal Mines Using LoRa Technology

Salna Joy[1], Ajay Sudhir Bale[1], M. I. Anju[2], R. Baby Chithra[1], Kavita Avinash Patil[4], and P. S. Neethu[3](✉)

[1] Department of ECE, New Horizon College of Engineering, Bengaluru 560103, India
[2] Department of ECE, SRM Institute of Science and Technology, Vadapalani Campus, Chennai 600026, India
[3] Department of ECE, School of Engineering and Technology, CHRIST (Deemed to be University), Bangalore, India
neethu.ps@christuniversity.in
[4] Division Electronics & Telecommunication Engineering, M.S. Ramaiah Institute of Technology, Bengaluru 560103, India

Abstract. Extracting coal from Underground mine is a hazardous and tough job that needs continuous monitoring of environmental conditions to protect workers' health and safety. Though some research works have explored wireless monitoring devices for underground mining, such as ZigBee and Wi-Fi technologies, they come with inherent restraints for instance restricted coverage, susceptibility to interference, reliability issues, security concerns, and high-power consumption. An Enhanced Safety Monitoring System for coal extraction from Underground Mines, employing LoRa communication technology for the effectual transmission of collected data to overcome existing challenges is discussed in this paper. The proposed system consists of two subsystems, one for monitoring the status of miners and another for comprehensive monitoring. LoRaWAN (Long Range Wide Area Network) is a multipoint protocol and this media access control (MAC) enables low-power devices to establish communication with Internet of Things (IoT) applications over extended wireless connections for long-range networks. LoRaWAN operates on lower radio frequencies, thereby providing a longer range of communication. This technology is known for its efficiency in optimizing LPWAN, offering extended range, extended battery life, robustness, and cost-effectiveness, making it highly suitable for industrial mining applications.

Keywords: LoRa Technology · LoRaWAN protocol · Internet of Things (IoT) · Underground Coal Mining · Low Power Wide Area Network (LPWAN)

1 Introduction

The mining sector, encompassing the extraction of coal, metals, and non-metals, holds substantial economic importance in numerous countries. These resources serve a multitude of applications and have a significant role in supplying vital components to various

V. E. Balas et al. (Eds.): IEQIP 2023, LNNS 1029, pp. 125–130, 2024.
https://doi.org/10.1007/978-3-031-61929-8_18

[1–3]. Coal, in particular, contributes to approximately 40% of the world's electricity generation and stands among the most significant sources of employment within this industry [4–6]. The growing demand for coal has led to an escalation in its production, which, in turn, has posed several challenges within mining operations [7–9]. These challenges include heightened methane emissions within mine workings and the release of carbon dioxide and carbon monoxide gases come out from the coal transportation facilities [10–12].

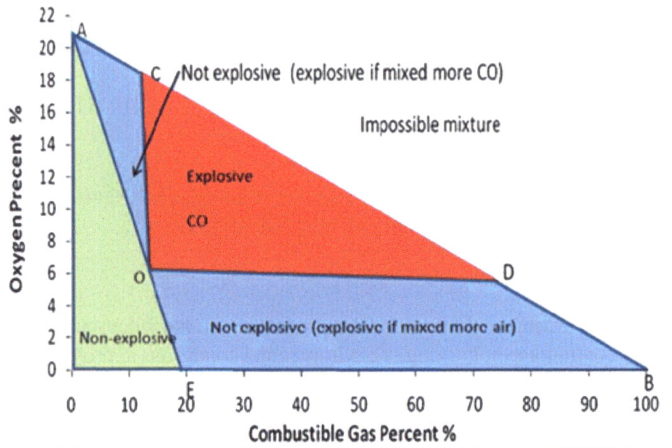

Fig. 1. Coward Explosive Triangle [9]

The diagram illustrating the CO explosive triangle is given in Fig. 1. While many existing monitoring systems in use for safety of miners rely on cable frameworks and wireless sensor frameworks, they are plagued by a host of issues. These problems include the inability to deliver essential information swiftly within the mines, restricted bandwidth, limited data transmission capabilities, challenging synchronization, and the vulnerability of sensors and cables to critical damage during accidents, particularly explosions. This damage hinders their capacity to relay information for search and rescue operations [13, 14]. To proactively address unforeseen incidents, there is a pressing need for a robust real-time gas monitoring system installed underground. Such a system should be capable of wireless communication by which data can be transmitted efficiently to the ground, thereby facilitating preventive measures [15, 16]. The primary aim of this paper is to create a system equipped with a wireless LoRa communication system for real time monitoring of the coal mines.

2 Proposed System Workflow

The system has 2 segments: one dedicated to monitoring the status of mine workers and the other focused on monitoring the condition of the mine unit. To keep track of unsafe gas levels, semiconductor gas sensors are employed. The gas sensors, Temperature sensor and a LoRa transmitter that are connected to the microcontroller are kept in the underground mines area. Pulse and MEMS sensors are attached to the workers.

For monitoring the health status of mine workers, Heartbeat and Respiratory sensors are deployed. In cases of emergency, a simple and easily accessible button is provided for immediate use. The MQ2 Gas Sensor Module is used to identify and notify users when there is an increase in gas levels. It converts the variation in gas concentration conductivity with regard to the output signal using a straightforward circuit. The MQ2 gas sensor is extremely sensitive. The permissible limits of different gases are given in Table 1.

Table 1. Safe Range of gas mixtures [16]

Sl. No	Gas Mixture	Safe Limit (%)
1	Methane	0.75(intake path)
		1.25(return path)
2	Carbon dioxide	0.5
3	Carbon Monoxide	0.005
4	Oxygen	19
5	Hydrogen Sulphide	0.0005

The monitoring system constantly measures environmental parameters and generates alarm when any of the Gas sensor value exceeds 500, Temperature exceeds 500C. Health monitoring subsystem stands for workers wellbeing and safety monitors pulse and respiration rate using pulse sensor and MEMS Respiratory sensor. In addition to being used for biofeedback applications like stress management and relaxation training, the respiratory sensors track the breathing in the abdomen and thoracic cavity. Additionally, the sensor reports on the relative depth of breathing.

The clothes can have the respiration sensors fastened to it. The majority of the respiratory sensor's placement is in the stomach area. To prevent tension loss, the sensor must be positioned tightly enough. When a finger is placed on a heartbeat sensor, the gadget is designed to produce digital outputs of the heartbeat. The microcontroller receives direct communication from these digital outputs in order to measure beats per minute (BPM). Alarm is generated if heartbeat falls below 60 bpm or crosses 100 bpm for pulse sensor output and MEMS sensor output falls outside the safe range of 12 and 25. Various sensors safe values are given in Table 2.

Table 2. Normal Range of Sensor values

Sl. No	Sensor	Normal Range
1	Gas Sensors	<500
2	Temperature sensor	<50
3	Pulse/Heartbeat Sensor	60 < value < 100
4	MEMS Sensor	12 < value < 25

3 Result and Discussion

The system successfully monitored various environmental parameters as well as health parameters. The Pulse rate, MEMS sensor output and gas sensor outputs measured over a period of two hours is plotted in Fig. 2, Fig. 3 and Fig. 4.

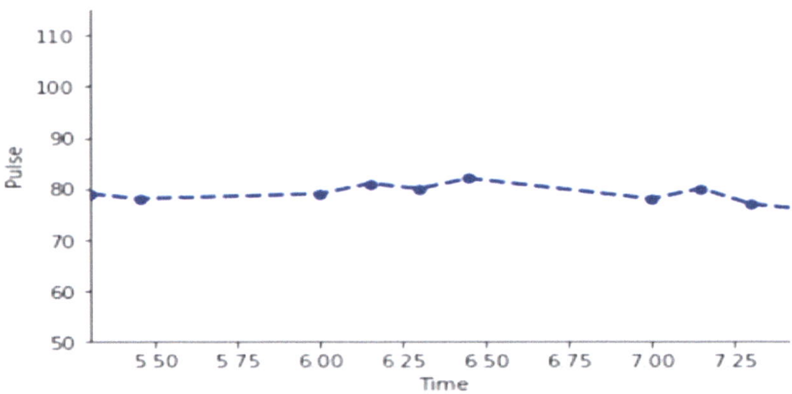

Fig. 2. Pulse Output

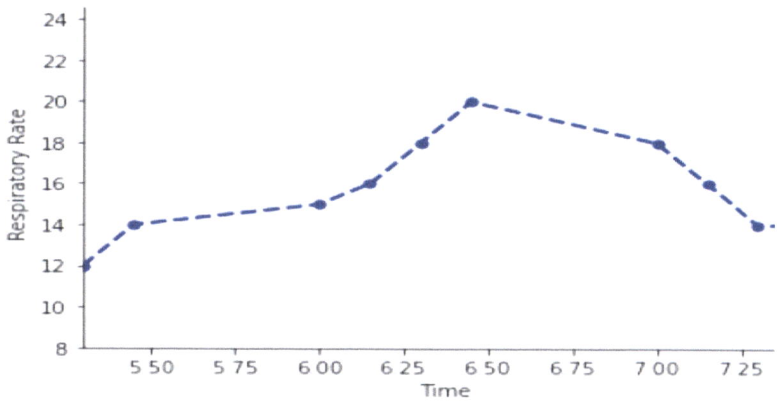

Fig. 3. MEMS Sensor Output

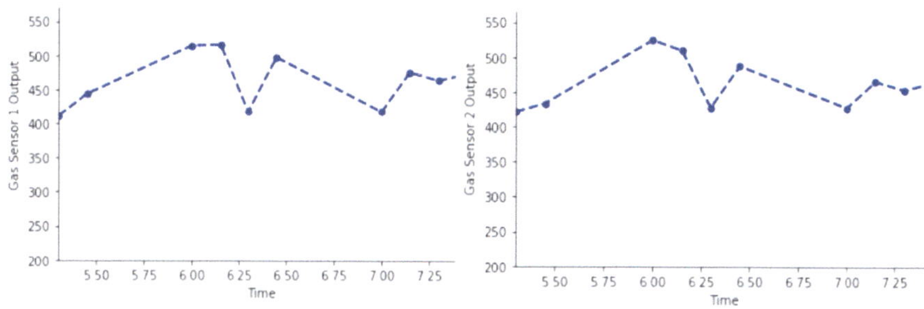

Fig. 4. Gas Sensor Outputs

4 Conclusion

A real-time monitoring system has been created to offer a more precise and location-specific view of conditions within the underground mine using LoRa communication devices connected through LoRaWAN gateway. The LoRa transmitter continuously sends the monitored data to the LoRa receiver located on the ground. The received data is subsequently uploaded to a webpage via the IoT connection. LoRa communication system provides enhanced signal strength and reduces the transmission lag between the Transmitter and Receiver. IoT implementation helps in Effective Real-Time Monitoring of the parameters and aids in the enhancement of the industrial conditions by evaluating parameter changes over time This technology serves as valuable support for miners, offering a proactive approach to safety, and allowing them to take preventive actions before any potential accidents occur. The system is designed to trigger an alarm when temperature or gas monitoring sensor readings exceed predefined thresholds. For monitoring the health status of mine workers, Heartbeat and Respiratory sensors are deployed. In cases of emergency, a simple and easily accessible button is provided for immediate use.

References

1. Kumar, P.P., Paul, P.S., Ananda, M.: Development of LoRa communication system for effective transmission of data from underground coal mines. Processes **11**(6), 1691 (2023). https://doi.org/10.3390/pr11061691
2. Zhang, H., Li, B., Karimi, M., Saydam, S., Hassan, M.: Recent advancements in IoT implementation for environmental, safety, and production monitoring in underground mines. IEEE Internet Things J. **10**(16), 14507–14526 (2023). https://doi.org/10.1109/JIOT.2023.3267828
3. Chen, W., Wang, X.: Coal mine safety intelligent monitoring based on wireless sensor network. IEEE Sens. J. **21**, 25465–25471 (2020)
4. Qiang, X., Li, G., Hou, J., Zhang, X., Liu, Y.: Intelligent safety risk analysis and decision-making system for underground metal mines based on big data. Sustainability (2023)
5. Khamis, Y., Alawi, M.A., Athumani, R., Sanya, W.: An IoT based worker safety helmet using cloud computing technology. Tanzania J. Eng. Technol. (2022)
6. Mishra, P.K., Swain, A., Kumar, S., et al.: Wireless paging system for underground mines. Radioelectron. Commun. Syst. **64**, 14–25 (2021). https://doi.org/10.3103/S0735272721010027
7. Ray Chowdhury, A., Pramanik, A., Roy, G.C.: IoT and LoRa based smart underground coal mine monitoring system. Microsyst. Technol. **29**, 919–938 (2023). https://doi.org/10.1007/s00542-023-05484-z
8. Branch, P.: Measurements and models of 915 MHz LoRa radio propagation in an underground gold mine. Sensors **22**(22), 8653 (2022)
9. Cheng, J., Luo, Y.: Modeling atmosphere composition and determining explosibility in a sealed coal mine volume. Arch. Min. Sci. **59**(1), 25–40 (2014)
10. Srivalli, G.: Study on coal mine safety monitoring and alerting system using IOT. Int. J. Sci. Technol. Eng. (2023). https://doi.org/10.22214/ijraset.2023.54194
11. Dhanalakshmi, V.: Keeping Track of Coal Mine Safety Using IoT Technology (2023). https://doi.org/10.1109/ICONSTEM56934.2023.10142538
12. Pratyush, K.: IoT Based Coal Mine Safety Monitoring and Alerting System (2022). https://doi.org/10.1109/ICAC3N56670.2022.10074169
13. Manasa, T., Kadali, J., Syed, N., Raju, G.S.V.S.K., Jamal, K.: IoT based coal mine safety monitoring and warning system. In: 2022 Sixth International Conference on I-SMAC (IoT in Social, Mobile, Analytics and Cloud) (I-SMAC), Dharan, Nepal, pp. 11–15 (2022). https://doi.org/10.1109/I-SMAC55078.2022.9987361
14. Erturk, M.A., Aydin, M.A., Buyukakkaslar, M.T., Evirgen, H.: A survey on LoRaWAN architecture, protocol and technologies. Future Internet **11**, 216 (2019)
15. Qiu, T., Chi, J., Zhou, X., Ning, Z., Atiquzzaman, M., Wu, D.O.: Edge computing in industrial Internet of Things: architecture, advances and challenges. IEEE Commun. Surv. Tutor. **22**(4), 2462–2488 (2020)
16. The Coal Mine Regulations (2017). https://www.dgms.gov.in/writereaddata/UploadFile/Coal_Mines_Regulation_1957.pdf. Accessed 27 Feb 2023

IoT Based Hydroponic System for Sustainable Organic Farming

Ajay Sudhir Bale[1](✉), K. A. Sharada[2], Sathya Vijaykumar[3], Naveen Ghorpade[4], Mamta B. Savadatti[1], and R. Hamsalekha[1]

[1] Department of ECE, New Horizon College of Engineering, Bengaluru 560103, India
ajaysudhirbale@gmail.com
[2] Department of CSE, HKBK College of Engineering, VTU, Bengaluru 560045, India
[3] Department of CSE, Atria Institute of Technology, Bengaluru 560024, India
[4] Department of CSE, Vijaya Vittala Institute of Technology, Bengaluru 562149, India

Abstract. Millions of people use medicinal plants as their only source of healthcare, making them a crucial global reference for illness therapy and the enhancement of health of people. Currently, excessive harvesting brought on by rising demand threatens the status of many therapeutic plants. Although numerous species are being grown in soil, the finished product frequently contains less medicinally active substances than plants which have been gathered from the wild. Studies shows that various kinds of medicinal plants have higher secondary metabolism content when cultivated hydroponically as opposed to when they are grown in the wild or in soil, which may lead to more sustainable access to these plants. There are numerous hydroponic systems and methods available, each with unique benefits and drawbacks. It is important to choose hydroponic systems that maximize the growth of the required substances and body parts since medicinal substances are frequently localized within particular plant organs. However, no attempt has yet been made to review the research with relation to the hydroponically grown farming of particular Parts of vegetation from medicinal plants. This paper examines the cultivation of medicinal plants using various hydroponic systems and how the system chosen varies based on the plant organ sought. It also highlights studies with cutting-edge findings that could advance this area.

Keywords: long-term viability, soilless growing · secondary metabolism · hydroponics

1 Introduction

Although agriculture constitutes one of among the most common activities worldwide, it is not consistent [1, 2]. An economy's vitality depends heavily on agriculture. It serves as the foundation of our economy. Various agricultural methods have been used to fulfil the demands of the growing population, including Plantation agriculture [3]: In this style of agriculture, the land is solely used to grow a single crop, such as cotton, coffee, or tea. Mixed agriculture [4–6]: This kind of farming combines the raising of animals with the production of crops [7, 8]. Poly house farming [9, 10]: In this method, crops are grown

V. E. Balas et al. (Eds.): IEQIP 2023, LNNS 1029, pp. 131–136, 2024.
https://doi.org/10.1007/978-3-031-61929-8_19

in a regulated environment. Innovative and alternative methods of food production are launched since they need a lot of land and labour, like: Aquaponics [11, 12]: This is the combination of hydroponics with aquaculture. Plants are grown hydroponically, or in water. Aeroponics [13, 14]: Growing plants air or in misty environment. In a hydroponic system, no soil is required at all; instead, a nutrient rich solution with a water base is utilized to produce plants, instead, materials like peat moss, clay pellets, perlite and rockwool supports the root of the plants. The technique employed keeps plants in good environmental condition by supplying all necessary nutrients in a water directly [15–18]. Because the nutrients are used effectively and are not squandered, the plants are able to grow properly as a result [19, 20].

2 Proposed Methodology

An IoT-based hydroponic system using Raspberry Pi Pico and programmed in Micro Python with LDR for lighting, temperature sensor, and pH sensor can be designed to provide optimal growing conditions for plants. The LDR sensor measures the light intensity, and the system adjusts the lighting system accordingly to ensure that the plants receive the right amount of light. As in Fig. 1, the submersible 230v AC motor is connected to power supply through 5v real and the motor is controlled by Arduino by giving time delay, which will pump water to reach the plant roots. The sensors are connected to controller, will monitor the intensity of light, temperature and purity of water.

The pH sensor in the hydroponic setup measures the acidity level of the nutrient solution, and the software running on the Raspberry Pi Pico continuously monitors the pH level. If the pH level falls below the desired range, indicating that the solution is too acidic, the software activates the chemical pump to add a base solution to the nutrient solution. This helps to increase the pH level and bring it back to the optimal range for plant growth.

The temperature sensor in the hydroponic setup measures the temperature, and the software running on the Raspberry Pi Pico continuously monitors the temperature level. If the temperature rises above the desired range, indicating that it is too high, the software activates the fan to blow cool air over the hydroponic setup. This helps to reduce the temperature and bring it back to the optimal range for plant growth.

The LDR sensor in the hydroponic setup measures the light intensity, and the software running on the Raspberry Pi Pico continuously monitors the LDR readings. If the light intensity falls below the desired range, indicating that the plants are not receiving enough light, the software activates the lighting system to increase the light intensity. If the light intensity rises above the desired range, indicating that the plants are receiving too much light, the software adjusts the lighting system to decrease the light intensity. This helps to maintain the optimal light conditions for plant growth.

Overall, these three features of the system work together to maintain optimal growing conditions for the plants in the hydroponic setup. By continuously monitoring the pH level, temperature, and light intensity and making adjustments as needed, the system can help ensure that the plants receive the right amount of nutrients, light, and temperature for healthy and productive growth. The Raspberry Pi Pico collects data from these sensors and sends it to the cloud-based platform Thing Speak, which processes the data and provides real-time visualization of the sensor readings.

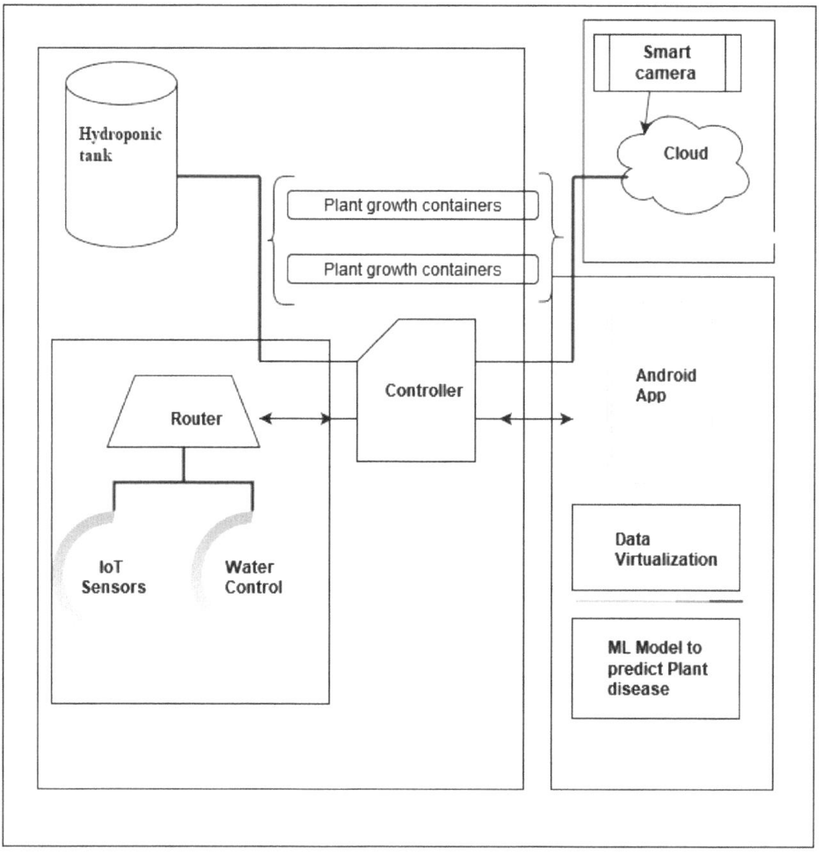

Fig. 1. Proposed Block diagram depicting the functions of this work

The pH sensor in the hydroponic setup measures the acidity level of the nutrient solution, and the software running on the Raspberry Pi Pico continuously monitors the pH level. If the pH level falls below the desired range, indicating that the solution is too acidic, the software activates the chemical pump to add a base solution to the nutrient solution. This helps to increase the pH level and bring it back to the optimal range for plant growth. Sustainable organic farming using hydroponic system as proposed in this work is shown in Fig. 1. The system has a power unit supply, router, microcontroller, hydroponic tank and container units for soil less plant growth. An android software unit aiding data visualization and predicting and reporting plant disease as and when needed, all these is done using ML models. Using the Cloud module in the circuit data is used to send the necessary messages to the owner alerting the health of the plant and growth sprouts, the images captured using camera/s. The Hydroponic System has a router and sensor unit; using a router gateway, data can be stored in a cloud server and sensor unit to monitor acidity, light intensity and temperature.

3 Results and Discussion

The Fig. 2 shows the overall model. Further the Light Intensity graph through LDR, Temperature control and Ph value is depicted in Fig. 3.

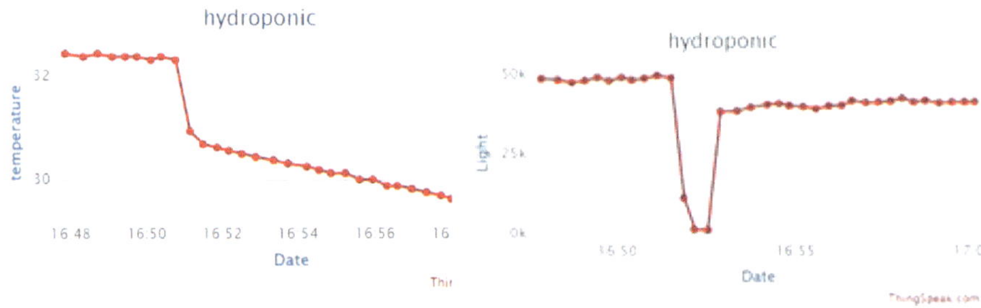

Fig. 2. Light intensity and temperature graph using Things Speak

The temperature, LDR, and pH graph for the IoT-based hydroponic system using Raspberry Pi Pico and programmed in Micro Python can provide valuable insights into the growing conditions of the plants. The temperature graph shows the temperature range of the hydroponic setup over time, allowing users to monitor and control the temperature of the setup to ensure that it remains within the optimal range for plant growth. The LDR graph shows the light intensity over time, which helps users to adjust the lighting system to provide the right amount of light for the plants. The pH graph shows the pH level of the nutrient solution over time, enabling users to maintain the pH level within the optimal range for plant growth. By analyzing these graphs, users can identify patterns and trends that may impact plant growth, such as temperature fluctuations or light intensity changes. Users can use this information to adjust the system's settings and customize control logic to optimize plant growth. Additionally, the real-time visualization of the sensor readings provided by Thing Speak can help users identify issues as soon as they occur, allowing them to respond quickly and prevent damage to the plants. Overall, the temperature, LDR, and pH graph is a powerful tool that can help users maintain healthy and thriving hydroponic plants.

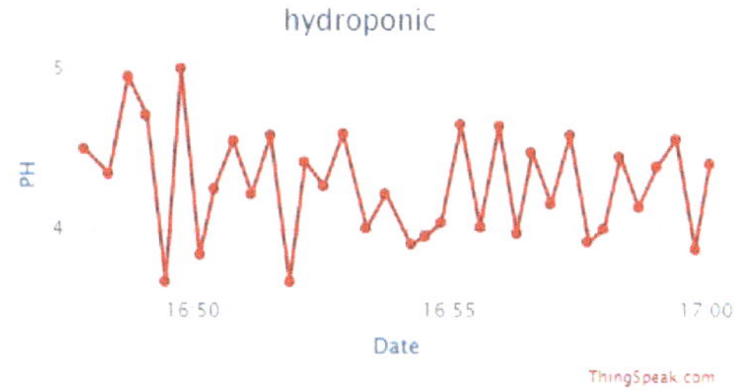

Fig. 3. PH value using Things speak

4 Conclusion

Hydroponic plants have a better bother resistance which dispenses with the higher utilize of pesticides. Compare to soil-based agriculture taste of this product is good. This strategy permits nourishment to be developed in regions that cannot bolster crops within the soil. It reduces the soil erosion as well as air and water pollution. This method of food production is much healthier than soil-based agriculture. This extend is effective in making completely controlled and programmed hydroponic framework with the assistance of ESP32 and all sensors. This frame work is exceptionally much appropriate in all locals independent of environment, which can be setup on. Based on the comparison between Smart Hydroponics and natural way of harvesting, Hydroponics gives more yields with less usage of water and pesticides by giving all nutrients to the plant.

References

1. Munandar, A., et al.: Design and development of an IoT-based smart hydroponic system. In: 2018 International Seminar on Research of Information Technology and Intelligent Systems (ISRITI), Yogyakarta, Indonesia, pp. 582–586 (2018). https://doi.org/10.1109/ISRITI.2018.8864340
2. Lu, N., Cheng, N., Zhang, N., Shen, X., Mark, J.: Connected vehicles: solutions and challenges. IEEE Internet Things J. **1**(4), 289–299 (2014)
3. Lakshmiprabha, K.E., Govindaraju, C.: Hydroponic-based smart irrigation system using Internet of Things. Int. J. Commun. Syst. **36**(12), e4071 (2023)
4. Kumara, V., et al.: Experimental study on a reliable smart hydroponics system. In: Human Agro-Energy Optimization for Business and Industry, pp. 27–45. IGI Global (2023)
5. Shrivastava, A., Nayak, C.K., Dilip, R., Samal, S.R., Rout, S., Ashfaque, S.M.: Automatic robotic system design and development for vertical hydroponic farming using IoT and big data analysis. Mater. Today: Proc. **80**, 3546–3553 (2023). https://doi.org/10.1016/j.matpr.2021.07.294
6. Shedek, M.A.A.B., Ali, R.A.M.: Smart hydroponic system using IoT. Evol. Electr. Electron. Eng. **4**(1), 287–294 (2023)
7. Dutta, M., et al.: Evaluation of growth responses of lettuce and energy efficiency of the substrate and smart hydroponics cropping system. Sensors **23**, 1875 (1875). https://doi.org/10.3390/s23041875
8. Venkatraman, M., Surendran, R.: Aquaponics and smart hydroponics systems water recirculation using machine learning. In: 2023 4th International Conference on Smart Electronics

and Communication (ICOSEC), Trichy, India, pp. 998–1004 (2023). https://doi.org/10.1109/ICOSEC58147.2023.10276310

9. Duarte, T., Gonçalves, P., Dias, G.P.: Low-cost smart hydroponic systems for home environment applications. In: 2023 18th Iberian Conference on Information Systems and Technologies (CISTI), Aveiro, Portugal, pp. 1–6 (2023). https://doi.org/10.23919/CISTI58278.2023.10211845
10. Srivani, P., Yamuna Devi, C.R., Manjula, S.H.: Prediction and comparative analysis using ensemble classifier model on leafy vegetable growth rates in DWC and NFT smart hydroponic system. In: Senjyu, T., Mahalle, P., Perumal, T., Joshi, A. (eds.) IOT with Smart Systems, pp. 795–804. Springer, Singapore (2022). https://doi.org/10.1007/978-981-16-3945-6_78
11. Wei, Y., Li, W., An, D., Li, D., Jiao, Y., Wei, Q.: Equipment and intelligent control system in aquaponics: a review. IEEE Access **7**, 169306–169326 (2019)
12. Blidariu, F., Grozea, A.: Increasing the economical efficiency and sustainability of indoor fish farming by means of aquaponics-review. Anim. Sci. Biotechnol. **44**, 1–8 (2011)
13. Dadi, P.S., Tamilvizhi, T., Surendran, R.: Layout optimization for agriculture or small scale agrarian industry. In: 2022 6th International Conference on Trends in Electronics and Informatics (ICOEI), pp. 174–179 (2022)
14. Gashgari, R., Alharbi, K., Mughrbil, K., Jan, A., Glolam, A.: Comparison between growing plants in hydroponic system and soil based system. In: Proceedings of the 4th World Congress on Mechanical, Chemical, and Material Engineering, Madrid, Spain, pp. 1–7. ICMIE (2018)
15. Kim, N., Ha, K.-J., Park, N.-W., Cho, J., Hong, S., Lee, Y.-W.: A comparison between major artificial intelligence models for crop yield prediction: case study of the midwestern United States, 2006–2015. ISPRS Int. J. Geo Inf. **8**(5), 240 (2019)
16. Rodríguez-Pérez, R., Vogt, M., Bajorath, J.: Support vector machine classification and regression prioritize different structural features for binary compound activity and potency value prediction. ACS Omega **2**(10), 6371–6379 (2017)
17. Ilyas, Q.M., Ahmad, M.: An enhanced ensemble diagnosis of cervical cancer: a pursuit of machine intelligence towards sustainable health. IEEE Access **9**, 12374–12388 (2021). https://doi.org/10.1109/ACCESS.2021.3049165
18. Ping, W.C., Salleh, M.F.M.: Wireless sensor network's agricultural monitoring system using TDMA based wireless communication. In: Mahyuddin, N.M., Noor, N.R.M., Sakim, H.A.M. (eds.) Robotics, Vision, Signal Processing and Power Applications: Enhancing Research and Innovation through the Fourth Industrial Revolution, pp. 955–962. Springer, Singapore (2022). https://doi.org/10.1007/978-981-16-8129-5_146
19. Saha, G.: Technological influences on monitoring and automation of the hydroponics system. In: 3rd IEEE International Virtual Conference on Innovations in Power and Advanced Computing Technologies i-PACT 2021 (2021)
20. Tagle, S., et al.: Development of an automated data acquisition system for hydroponic farming. In: 2018 IEEE 10th International Conference on Humanoid Nanotechnology Information Technology Communication and Control Environment and Management, HNICEM 2018 (2019)

Predicting Stride Length from Acceleration Signals Using Lightweight Machine Learning Algorithms

To-Hieu Dao[1,2,3], Duc-Nghia Tran[4], Van-An Tran[1], Hoang-Dieu Vu[1,2,3], Kolla Bhanu Prakash[5], and Duc-Tan Tran[1](✉)

[1] Phenikaa University, Yen Nghia, Hanoi, Vietnam
{Hieu.daoto,dieu.vuhoang,tan.tranduc}@phenikaa-uni.edu.vn
[2] PRATI, A&A Green Phoenix Group JSC, Cau Giay, Hanoi, Vietnam
[3] Graduate University of Sciences and Technology, VAST, Hanoi, Vietnam
[4] Institute of Information Technology, VAST, Hanoi, Vietnam
nghiatd@ioit.ac.vn
[5] Koneru Lakshmaiah Education Foundation, Vaddeswaram, Andhra Pradesh, India

Abstract. In the field of motion analysis and behavioral movement research, accurately measuring and predicting stride length from acceleration signals plays a critical role. Recent studies have undertaken the development and evaluation of numerous machine learning models to address this issue, focusing on the deployment of lightweight and efficient models on devices with limited computational capabilities, such as microcontrollers. Among the algorithms examined, random forest, k-nearest neighbors, support vector machines, decision trees, and extreme gradient boosting have been selected due to their balance between prediction accuracy and computational lightness. These algorithms not only ensure a high degree of accuracy in predicting stride length from collected acceleration data but also ensure that the models can be deployed directly on smart wearable devices and microcontrollers, extending the practical application of motion monitoring systems. The outcomes of these machine learning models promise to contribute to the development of personalized health monitoring solutions, as well as in applications related to sports and functional rehabilitation, where analyzing and improving gait is extremely important.

Keywords: Acceleration · Machine Learning · Microcontrollers · Stride Length · Prediction

1 Introduction

In the past, advancements in sensor technology and machine algorithms opened up numerous potential research directions in monitoring and analyzing human actions [1–6]. Among these, estimating step length using accelerometers has become a research topic of burgeoning interest due to its wide applications in healthcare, sports, security, as well as the challenges posed in processing and analyzing motion data [7–13]. Recent

V. E. Balas et al. (Eds.): IEQIP 2023, LNNS 1029, pp. 137–143, 2024.
https://doi.org/10.1007/978-3-031-61929-8_20

studies have focused on developing lightweight machine learning algorithms capable of running directly on wearable devices [1, 3–6, 9], aiming to provide real-time predictive information to users. The combination of lightweight algorithms and accelerometers not only improves energy efficiency but also reduces limitations in computational speed with cloud data and central computing resources [3, 5]. However, accurately modeling human motion remains a significant challenge, especially in differentiating between types of actions and dealing with the diversity of individual walking styles [10–12]. Algorithms need to be flexible enough to adapt to minor variations in signals without compromising the accuracy of the overall predictions. This requires continuous improvements in both data collection and machine learning techniques, particularly the degree to which prediction models correlate with actual action data [14, 15]. This research proposed the application of lightweight machine learning algorithms in processing accelerometer data in real-time as humans walk and take steps. A stride (Fig. 1) was understood as the long stepping process of one leg, typically comprising two stages: i/A short preparatory step made by one leg (step-up); ii/Followed by the other leg taking a long step (stride) to move the body forward.

2 Research Methods

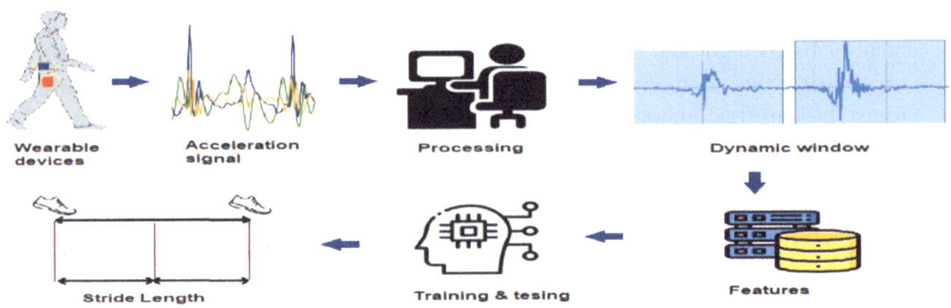

Fig. 1. The process of building a system model estimates the length of a stride.

The architecture and the strides to construct the model predicting the stride length are illustrated in Fig. 1. The process includes five main steps: data collection, preprocessing, sliding window, features computation, training & testing. Specifically as follows:

Data Collection. Data on the stepping process was collected by the acceleration signal from sensors integrated on wearable devices attached to the user's body (smartphone [16], smartwatch [4], data collection devices [1, 5, 6]. To construct and evaluate the model, this study used a publicly shared dataset on stride length estimation at [7]. This study evaluated the model based on accelerometer data collected from two positions: pelvis and thigh areas, places that are less likely to cause discomfort to the wearer while moving and are minimally affected by actions unrelated to the step motion. The total number of stride length windows that were obtained was 49,862 (Table 2). To ensure that the data segmentation was suitable for the construction of the stride length learning model and to improve the quality of the predictive model, 1/3 of the stride length data from each volunteer was chosen to create the training set, and the remaining 2/3 was used for the testing set (Table 1).

Table 1. The dataset consists of step motion data collected from 10 volunteers who moved at a constant speed. The number of strides taken at each speed (slow, normal, fast) and by each volunteer varies.

Person	Height (cm)	Leg (cm)	The number of strides					
			Thigh region			Pelvis region		
			Slow	Nomarl	Fast	Slow	Nomarl	Fast
Person01	1.81	1.09	875	884	923	875	884	923
Person02	1.88	1.10	735	852	886	735	852	886
Person03	1.88	1.16	675	756	862	0	756	862
Person04	1.81	1.09	759	844	941	759	844	0
Person05	1.80	1.07	755	842	935	755	842	935
Person06	1.67	1.02	808	838	911	808	838	911
Person07	1.60	0.9	747	851	1151	747	851	1151
Person08	1.69	1.02	867	892	907	867	892	907
Person09	1.70	1.03	810	872	1077	810	872	1077
Person10	1.83	1.05	729	823	932	729	823	932

Pre-processing. To concentrate on the motion information from the three axes of acceleration (x_i, y_i, and z_i), this study utilized motion data on the y-axis and calculated the angular variation of the device $\emptyset_i$ over time based on the orientation of the y-axis y_i (Eq. 1).

$$\emptyset_i = \arctan\left(\frac{y_i}{\sqrt{x_i^2 + z_i^2}}\right) \tag{1}$$

Sliding Window. The sliding window technique [1] allows for the analysis and detection of the stride region from the moment the foot is lifted to the moment it is placed down, significantly improving the ability to identify and measure the stride length.

Features Computation. After the data was segmented, features (Table 2) were extracted on each corresponding data window.

Training and Testing. Aiming to build real-time predictive models for stride length based on accelerometer sensors, we developed predictive models using 5 machine learning regression algorithms including: decision tree (DT), random forests (RF), extreme gradient boosting (XGboost), support vector machine (SVM), and k-nearest neighbors (KNN) với K = 5.

Table 2. The features extracted on each window after the data has been segmented and labeled

Features	Description	Equation	
Mean	The mean value of the signal series, calculated by summing the values and dividing by the number of values	$\mu = \frac{1}{n}\sum_{i=1}^{n} x_i$	(2)
Standard deviation	The square root of the average of the squared distances from each data point to the mean	$\sigma = \sqrt{\frac{1}{n}\sum_{i=1}^{n}(x_i - \mu)^2}$	(3)
Root mean square	The square root of the average of the squares of the values	$RMS = \sqrt{\frac{1}{n}\sum_{i=1}^{n} x_i^2}$	(4)
Range	The difference between the largest and smallest values in the dataset	$Range = Max(x_i) - Min(x_i)$	(5)
Entropy	Entropy of the signal frequency on the y-axis	$S(X) = -\sum_1^n p(x_i) * log_2(p(x_i))$	(6)
Height	Volunteer's height		
Leg length	The length of the volunteer's calf		

3 Model Evaluation

The predicted models were evaluated based on three parameters: R-squared (Eq. 7), Mean Absolute Error (Eq. 8) and standard deviation. In there, $\hat{y}_i$ is predicted values, y_i is true values and $\overline{y_i}$ is the average value.

$$R^2 = 1 - \frac{\sum_{i=1}^{n}\left(y_i - \widehat{y_i}\right)^2}{\sum_{i=1}^{n}\left(y_i - \overline{y_i}\right)^2} \tag{7}$$

$$MAE = \frac{1}{n}\sum_{i=1}^{n}\left|y_i - \widehat{y_i}\right| \tag{8}$$

4 Results and Discussion

The evaluation results for the prediction models DT, Xgboost, SVM, KNN, and RF were tabulated in Table 3. The Random Forest model had the highest performance among the five models tested. Specifically, it stood out with the highest R2 value of 0.68, indicating the best overall explanatory ability for the data variability. Furthermore, for different walking speeds and wearable device positions, RF consistently maintained an MAE below 6 cm and R2 usually above 0.5, proving its stability and reliability for this prediction task. Although XGBoost did not perform as well as Random Forest, it provided reasonable results with average MAE and R2 values. This model's MAE ranged from 6.580 cm to 6.957 cm, with the lowest R2 at 0.467 and the highest at 0.55. A significant difference emerged with the SVM model, whose instability was evident through the large fluctuations in MAE and R2 across different walking speeds. For thigh placement, the highest MAE was 8.182 cm, and the lowest was 7.277 cm, with R2 varying between 0.2 and 0.385, indicating SVM's inconsistency across different test conditions for wearable positions and speeds. Additionally, the KNN (k = 5) model showed the lowest performance with the highest MAE and lowest R2 among the five models. With the R2 index was 0.068 và MAE at 0.068 and an MAE of up to 7.986 cm, this model was the least suitable for prediction in this context. A similar outcome was observed with the DT model, whose performance was uneven. For thigh placement, MAE varied from 7.463 cm (fast speed) to 9.45 cm (slow speed), and R2 from 0.341 (fast speed) to 0.609 (slow speed). Overall, the prediction models tended to perform better at the pelvis position compared to the thigh. This might be due to differences in data variability and the impact of the stepping process between the two body parts, with the pelvis possibly having less variability in motion compared to the thigh. Movement speed was also an important factor, but no model showed clear differences in performance metrics at different speeds.

Table 3. The evaluation metrics used when the regressor models were applied

Model	Index	Speed at Thigh			Speed at Pelvis		
		Low	Nomarl	Fast	Low	Nomarl	Fast
DT	MAE (cm)	9.45	7.463	7.483	6.953	7.856	7.76
	SD (cm)	10.3	6.439	6.430	6.276	6.699	6.566
	R^2	0.609	0.341	0.329	0.411	0.278	0.227
XGboost	MAE (cm)	6.623	6.684	6.593	6.580	6.957	6.694
	SD (cm)	4.908	5.223	4.723	5.475	5.495	4.852
	R^2	0.550	0.512	0.547	0.508	0.467	0.489
SVM	MAE (cm)	7.332	7.382	7.624	7.603	8.182	7.961
	SD (cm)	6.611	6.018	6.212	6.236	6.346	6.598
	R^2	0.356	0.385	0.334	0.351	0.273	0.200
KNN	MAE (cm)	7.779	7.93	7.277	7.986	8.006	8.549
	SD (cm)	6.692	6.823	6.302	7.324	6.482	7.175
	R^2	0.304	0.258	0.362	0.211	0.281	0.068
RF	MAE (cm)	5.373	5.866	5.782	5.38	6.11	5.94
	SD (cm)	4.445	4.585	4.512	4.57	4.82	4.63
	R^2	0.68	0.62	0.63	0.67	0.59	0.58

5 Conclusion

The application of dynamic time segmentation with varying window sizes allowed for improving the speed of detecting the time intervals during stride occurrences and focusing on the necessary information for prediction models. Research results indicated that random forest and extreme gradient boosting models yielded good results in constructing models for predicting step length using accelerometer sensors. In contrast, decision tree models, support vector machines, and k-nearest neighbors were found to be less effective and slower in adapting to data variations describing step length. The predictions of step length at different walking speeds demonstrated that step length was not dependent solely on a person's walking speed but was influenced by various factors such as constant walking speed, height, and leg length. In this research, a prediction model based on the random forest algorithm achieved good results, with the lowest MAE recorded at 4.73 cm, while SD and R^2 remained stable at approximately 4.5 cm and 0.6, respectively. In the future, we tend to develop prediction models capable of adapting to free walking and varying speed movements, aiming to apply them to more complex tasks such as locating firefighters, monitoring the walking process of the elderly, and so on. This will require optimized systems and applications that are not only compact and comfortable to wear but also integrate affordable, energy-efficient sensors while maintaining high performance.

Acknowledgment. This research is funded by Phenikaa University under grant number PU2023-1-A-07.

References

1. Dao, T.-H., Tran, D.-N., Hoang, Q.-T., Vu, H.-D., Huy, D. T., Tran, D.-T.: Developing real-time automatic step detection on a low-cost, performance-constrained microcontroller. In: 2023 IEEE Statistical Signal Processing Workshop, pp. 150–154 (2023)
2. Vu, H.-D., Tran, D.-N., Can, K.-L., Dao, T.-H., Pham, D.-D., Tran, D.-T.: Enhancing sleep postures classification by incorporating acceleration sensor and LSTM model. In: 2023 IEEE Statistical Signal Processing Workshop (SSP), pp. 661–665 (2023)
3. Tran Thi Hong, N., Nguyen, G.L., Quang Huy, N., Viet Manh, D., Tran, D.-N., Tran, D.-T.: A low-cost real-time IoT human activity recognition system based on wearable sensor and the supervised learning algorithms. Measurement **218**, 113231 (2023)
4. Minh, N.C., Dao, T.H., Tran, D.N., Nguyen, Q.H., Nguyen, T.T., Tran, D.T.: Evaluation of smartphone and smartwatch accelerometer data in activity classification. In: 2021 8th NAFOSTED Conference on Information and Computer Science (NICS), pp. 33–38 (2021)
5. Dao, T.H., Yen, H.T.H., Hoang, V.N., Tran, D.T., Tran, D.N.: Human activity recognition system for moderate performance microcontroller using accelerometer data and random forest algorithm. EAI Endorsed Trans. Ind. Netw. Intell. Syst. **9**(4), 1–18 (2022). https://doi.org/10.4108/eetinis.v9i4.2571
6. Thu, N.T., Dao, T., Bao, B.Q., Tran, D., Van Thanh, P., Tran, D.-T.: Real-time wearable-device based activity recognition using machine learning methods. Int. J. Comput. Digit. Syst. **12**(1), 321–333 (2022)

7. Vezočnik, M.: Benchmark dataset for performance evaluation of inertial sensor-based step length estimation models (2021). https://github.com/repositoryadmin/SLERepository/tree/master/SLEDataset2. Accessed 09 Nov 2023
8. Shin, S.H., Park, C.G., Kim, J.W., Hong, H.S., Lee, J.M.: Adaptive step length estimation algorithm using low-cost MEMS inertial sensors. In: Proceedings of the 2007 IEEE Sensors Applications Symposium, SAS, no. February, pp. 6–8 (2007). https://doi.org/10.1109/SAS.2007.374406
9. Seo, K.: Real-time estimation of walking speed and stride length using an IMU embedded in a robotic hip exoskeleton. In: Proceedings of the IEEE International Conference on Robotics and Automation, vol. 2023-May, no. ICRA, pp. 12665–12671 (2023). https://doi.org/10.1109/ICRA48891.2023.10160770
10. Kim, J.W., Jang, H.J., Hwang, D.-H., Park, C.: A step, stride and heading determination for the pedestrian navigation system. J. Glob. Position. Syst. **3**(1&2), 273–279 (2004). https://doi.org/10.5081/jgps.3.1.273
11. Vezocnik, M., Juric, M.B.: Average step length estimation models' evaluation using inertial sensors: a review. IEEE Sens. J. **19**(2), 396–403 (2019)
12. Vezočnik, M., Juric, M.B.: Adaptive inertial sensor-based step length estimation model. Sensors **22**(23) (2022). https://doi.org/10.3390/s22239452
13. Xu, G., Wu, Q., Daneshmand, M., Liu, Y.: A data privacy protective mechanism for WBAN. Wirel. Commun. Mob. Comput. 421–430 (2015). https://doi.org/10.1002/wcm
14. Pham, V.T., et al.: Highly accurate step counting at various walking states using low-cost inertial measurement unit support indoor positioning system. Sensors **18**(10) (2018). https://doi.org/10.3390/s18103186
15. Van Thanh, P., Thi, A.-D. N., Thuy, Q.T.T., Phuong, D.C.T., Mau, V.H., Tran, D.-T.: A novel step counter supporting for indoor positioning based on inertial measurement unit. In: 2017 7th International Conference on Integrated Circuits, Design, and Verification (ICDV), pp. 69–74 (2017). https://doi.org/10.1109/ICDV.2017.8188641
16. Kwapisz, J.R., Weiss, G.M., Moore, S.A.: Activity recognition using cell phone accelerometers. ACM SIGKDD Explor. Newsl. **12**(2), 74–82 (2011)

Unveiling Hate: Multimodal Perspectives and Knowledge Graphs

Archana Lopes(✉), David Jennicson, Kunal Wagh, and Akash Agre

Fr. Conceicao Rodrigues College of Engineering, Bandra, Mumbai, India
archana_lopes@frcrce.ac.in

Abstract. Deep learning models continue to face challenges due to the growing amount of hate speech spread online. The recent advances in research have led to the development of several models that may be utilised to address these difficulties. This covers both unimodal and multimodal models; however, due to accuracy limits and performance restrictions, a more sophisticated method is required for these kinds of issues. This study evaluates the effectiveness of previously proposed models and frameworks for the identification of hateful memes and makes recommendations for how to improve them to increase accuracy. The significance of multimodal techniques for categorising multimodal content—like memes, which include text and picture features—is also highlighted in the research. Knowledge graphs are used in the proposed approach to solve the problem of damaging analogies and stereotypes found in memes. External world information is also used to create multimodal models that effectively categorise memes that are closely connected to real-world events and data.

Keywords: Multimodality · Sentimental analysis · Ensemble learning · Knowledge graph · Transformer models

1 Introduction

The internet and social media fuel the spread of hateful speech, often through memes. Deep learning models struggle to classify memes due to their multimodal nature. Memes, combining various elements, pose a challenge for accurate classification. To determine if a meme is hateful, one must consider its context. Hateful memes target and demean individuals or groups based on personal characteristics. Multimodal approaches include early fusion, combining modalities before decision-making, and late fusion, processing features independently. Early fusion models outperform late fusion ones in handling complex multimodalities. Existing research favours early fusion but neglects current scenarios, such as evolving meme subjects. To address this, the paper proposes the usage of knowledge graphs to integrate analogies in word embeddings. This approach identifies harmful analogies, like associating a race with criminal behaviour. The knowledge graph aids in analyzing and mitigating negative impacts effectively.

V. E. Balas et al. (Eds.): IEQIP 2023, LNNS 1029, pp. 144–149, 2024.
https://doi.org/10.1007/978-3-031-61929-8_21

2 Related Work

This research is motivated by recent advancements in deep learning algorithms for natural language processing (NLP) and image feature extraction. The paper aims to enhance our understanding of image feature extraction, segmentation, object recognition, and NLP for text production through experimentation and accuracy testing of various models and methods.

There has been a significant increase in scientific interest in identifying offensive memes in recent years. Various approaches have been developed to address this issue, using different methods. One common approach involves text and feature extraction, followed by vector representation and classification. Ensemble models that combine several models [1–4] are a popular strategy for accurate predictions. To improve the identification of offensive memes, certain algorithms, such as Visual-BERT [5], Vil-BERT [6], Ernie-Ville [7], UNITER-VILLA [8], and OSCAR [9], integrate both textual and visual data. These algorithms use early and late fusion techniques to combine data from multiple modalities, and they have proven to be effective in identifying offensive memes.

Research on knowledge graphs to address negative stereotypes in content, such as memes, includes studies by Zi Ye et al. [10] using Graph Neural Networks, Poojitha Maheshappa et al. [11] using LSTM networks with DBPedia knowledge graph embeddings for hate speech detection, and Pushkar et al. [12] employing Graph Convolutional Networks for abusive language detection. These approaches showcase the potential of knowledge graphs and related strategies in identifying and mitigating negative stereotypes in multimodal content.

3 Proposed Methodology

3.1 Introduction

To identify memes, the initial step involves extracting traits from both text and visual elements. Memes, often conveying humour like sarcasm or satire, pose a challenge due to diverse contexts and potential offensiveness. Accurate identification necessitates a comprehensive approach, including feature extraction, contextual understanding, and semantic analysis. The study references the Hateful Meme Detection Challenge[13] dataset from Meta AI, comprising 10,000 photos—8500 for training, 500 for validation, and 1,000 for testing.

3.2 Multimodal Extraction

The approach is similar to the methodology outlined by Ron Zhu [1], who achieved the highest accuracy in the Hateful Meme Detection Challenge. However, this approach relies on a third-party solution, despite the goal of creating an open-source solution. Critical stages include object detection, text extraction, picture captioning, and group learning. Unlike Zhu's approach, this method emphasizes extracting facial features like gender, age, and ethnicity for decision-making. Various techniques were explored for face attribute extraction (Fig. 1).

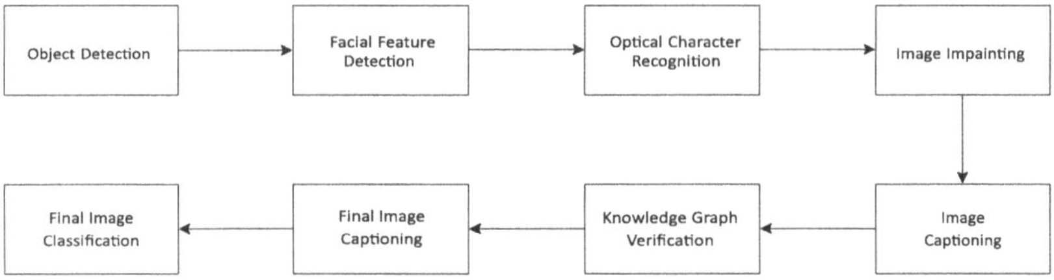

Fig. 1. Flowchart for the Process

Initially using YOLO v8 with the FairFace dataset [14] faced challenges, leading to the choice of Deepface [15] with 97.5% facial recognition accuracy. Deepfilv2, a gated convolutions-based approach [16], restored images by inpainting text-masked areas. Prior investigations inform text and facial feature extraction and object identification. The approach involves a comprehensive review and analysis of current techniques, suggesting new methodologies and tools for an ideal and refined process.

3.3 Knowledge Graphs

Various approaches address stereotyped analogies in online content. Utilizing knowledge graphs such as DBpedia or creating a new network by aggregating reliable data from sources like Wikipedia is a useful approach. The knowledge graph uses Neo4j to store extracted entities and relationships and is designed to encompass all possible stereotyped comparisons. During operation, entities from multimodal models are queried to identify links between them, aiding effective information classification. Each entity forms a node, interconnected by edges in the knowledge graph. The objective is to gather, generalize, and convert data from diverse sources into entities and relationships. Additionally, the approach integrates stereotyped analogies from previous research and literature into the knowledge base, making it more comprehensive and effective.

3.4 Classification

The evaluation of various existing methodologies is intended in the context of meme categorization. The evaluation includes the use of Supervised machine learning methods such as logistic regression, Random Forest, XGBoost, and NLP unimodal models like BERT. Although previous research has shown that these models perform less well than ideal, it is expected that using a knowledge graph would significantly improve the results. The approach used makes it easier to find connections in memes efficiently, even when doing so requires knowledge of the outside world, which was a problem with earlier methods. The goal is to address this shortcoming and improve the standard of results shown in previous attempts.

4 Results and Discussions

During the detection of objects in the given meme image using YOLOv8, the results were positive. The deepface model was used for processing specialized data, specifically for classifying human face racial features. Additionally, the addition of the face_recognition library played a crucial role in obtaining information about famous personalities and improving knowledge extraction from graphs. Table 1 and 2 provides a brief overview of the performance of OCR models in extracting text from memes. Easy-OCR effectively identified text regions in photos despite a high Character Error Rate (CER) and varied font styles in meme images. The transformer-based OCR (Tr-OCR)[17] model, trained on diverse font-type datasets, proved helpful in addressing this challenge. Mistakes decreased by narrowing the input to only the text portions of the meme image, leading to significant improvements in accuracy, CER, and WER errors. Future research and development will prioritize this area (Fig. 2).

Fig. 2. The input image fed into OCR.

Table 1. OCR Tool Results

OCR Tool	Input Image	Output	Result
EasyOCR	Whole Meme Image	'@hem vou {orget @0'	0.7827
EasyOCR	Cropped image	[lhen wou forgjet t'	0.92406
TrOCR	Whole meme image	':'	0.000
TrOCR	Cropped image	'WHEN YOU FORGET'	1.0

The coco-large version of the Generative Image Transformer (GIT) model is employed for image captioning, chosen for its excellent performance on the available hardware—a Ryzen 5 5600H CPU and a GTX 1650 graphics card. Inference time has been extended to meet project requirements, and the model adeptly identifies objects, people, and locations, fulfilling all project criteria. In addressing stereotypical analogies in online content, ongoing research explores various strategies. This includes using established knowledge graphs like DBpedia or creating a new graph with data from reliable sources such as news outlets and Wikipedia. The knowledge base aims to incorporate potential stereotypical analogies, driving the graph's creation. Entities and relationships are extracted and stored in a sophisticated graph database like Neo4j, essential for search queries to extract context from a meme. The knowledge graph operates by searching for entities from multimodal models and querying relationships. This enhances efficiency in the classification process. Nodes represent entities, and edges denote relationships. Information from diverse sources is generalized into entities and relationships, with a plan to integrate stereotypical analogies from existing literature and research studies. The research entailed fine-tuning the BERT model through the utilization of the Siamese BERT architecture. This architecture incorporates multiple inputs and produces a single binary output that signifies class labels (0 for non-offensive and 1 for offensive). The model underwent training for a total of 3 epochs, and the outcomes are outlined below (Fig. 3):

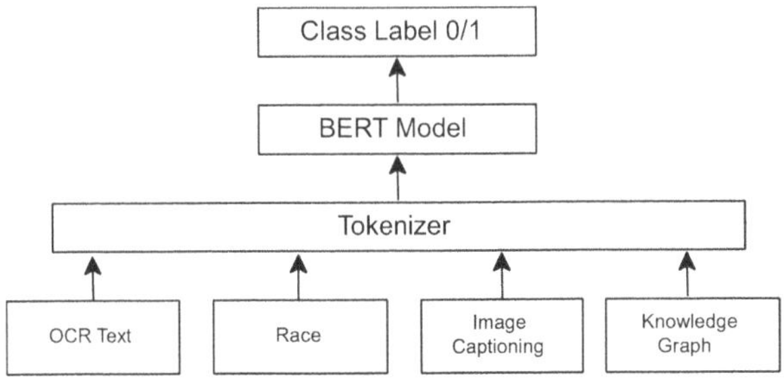

Fig. 3. The Siamese BERT model

Table 2. Total number of epoch and respective accuracies

Epoch	Training Loss	Validation Accuracy
1/3	0.6091	0.7120
2/3	0.5234	0.7193
3/3	0.4430	0.7230

Experiments using an encoder-decoder model, specifically the T5 model, aimed to classify and generate textual justifications for meme images. Training on a small dataset of 8209 photos revealed limited effectiveness, attributed to the dataset's size. Five cross-validation cycles were performed, indicating room for improvement in model accuracy based on observations.

5 Conclusion

The paper explores the potential of using knowledge graphs, which have shown promising results. The proposed solution is a knowledge graph that is based on plain text rather than vector-based embedding. With the incorporation of vector-based text embeddings and image search, subsequent endeavours could significantly expand the scope of this project, aiming to achieve optimal accuracy.

References

1. Zhu, R.: Enhance multimodal transformer with external label and in-domain pretrain: hateful meme challenge winning solution (2020)
2. Muennighoff, N.: Vilio: state-of-the-art visio-linguistic models applied to hateful memes (2020)

3. Velioglu, R., Rose, J.: Detecting hate speech in memes using multimodal deep learning approaches: prizewinning solution to hateful memes challenge (2020)
4. Lippe, P., et al.: A multimodal framework for the detection of hateful memes (2020)
5. Li, L.H., Yatskar, M., Yin, D., Hsieh, C.-J., Chang, K-W.: VisualBERT: a simple and performant baseline for vision and language (2019)
6. Lu, J., Batra, D., Parikh, D., Lee, S.: ViLBERT: pretraining taskagnostic visiolinguistic representations for vision-and-language tasks (2019)
7. Chen, Y.-C., et al.: UNITER: universal image-text representation learning (2020)
8. Feng, Z., et al.: ERNIE-ViLG 2.0: improving text-to-image diffusion model with knowledge-enhanced mixture-of-denoising-experts (2022)
9. Li, X., et al.: Oscar: objectsemantics aligned pre-training for vision-language tasks (2020)
10. Ye, Z., Kumar, Y.J., Sing, G.O., Song, F., Wang, J.: A comprehensive survey of graph neural networks for knowledge graphs (2022)
11. Maheshappa, P., Mathew, B., Saha, P.: Using knowledge graphs to improve hate speech detection (2021)
12. Mishra, P., Tredici, M.D., Yannakoudakis, H., Shutova, E.: Abusive language detection with graph convolutional networks (2019)
13. Kiela, D., et al.: The hateful memes challenge: detecting hate speech in multimodal memes (2020)
14. Karkk, K., Joo, J.: FairFace: face attribute dataset for balanced race, gender, and age for bias measurement and mitigation (2021)
15. Taigman, Y., Yang, M., Ranzato, M.A., Wolf, L.: DeepFace: closing the gap to human-level performance in face verification (2014)
16. Yu, J., Lin, Z., Yang, J., Shen, X., Lu, X., Huang, T.: FreeForm image inpainting with gated convolution (2019)
17. Li, M., et al.: TrOCR: transformer-based optical character recognition with pre-trained models (2022)
18. Sandulescu, V.: Detecting hateful memes using a multimodal deep ensemble (2020)
19. Aggarwal, P., et al.: HateProof: are hateful meme detection systems robust? (2023)

Vision-Based Toddler Activity Recognition: Challenges and Applications

P. Shilpa(✉) and M. Chandrajit

Maharaja Research Foundation, Maharaja Institute of Technology Mysore, Mysuru, Karnataka, India

Shilpa628@gmail.com, chandrajithmmca@mitmysore.in

Abstract. Advances in Artificial Intelligence and Computer Vision have made automation systems popular in a variety of fields. One of these is Human Activity Recognition (HAR), because the HAR is a complex system, many issues and challenges need to be addressed. One of the most challenging research areas in HAR is identifying toddler activities from various vision input devices. In order to enhance the safety and wellness of toddlers, their behavior has to be studied separately. This research examines existing vision-based toddler activity recognition systems, available datasets to conduct experiments, challenges involved, and potential societal applications. Furthermore, some interesting problems and ideas for future research are presented.

Keywords: Automated systems · Computer Vision · Human Activity Recognition · Toddler

1 Introduction

Human Activity Recognition (HAR) is the process of identifying and naming the activity performed by humans such as sitting, standing, walking, running, the interaction between two or more people, and the interaction between humans and objects [1–3], etc. Some of the applications of HAR include Human-Computer Interaction (HCI), Ambient Assisted Living (AAL), intelligent video surveillance, human behavior understanding, robotic learning, virtual reality, home monitoring, crowd analysis and health care. Based on data acquisition, HAR is divided into sensor-based HAR and computer vision-based HAR [4]. Video analysis tasks have been moving from inferring the present state to predicting the future state. The complexity and diversity of human activities make it difficult to identify and classify human activities from visual sensory devices. Identification of activity in the scene requires temporal identification and localization [2, 4]. One of the most challenging research areas under HAR is computer vision-based toddler activity recognition.

Vision-based Toddler Activity Recognition (VTAR) is a special class of HAR that uses visual data collected from various visual input devices to detect and forecast toddler activity [5]. The activities of a toddler are climbing stairs, crawling, playing with toys,

V. E. Balas et al. (Eds.): IEQIP 2023, LNNS 1029, pp. 150–156, 2024.
https://doi.org/10.1007/978-3-031-61929-8_22

riding a bicycle, walking downstairs, and yawning to name a few. This study aims to examine toddler activity from birth to three years of age.

There are a number of reasons why toddlers' activities should be examined. They include clinical purposes, including detecting seizure-like behavior, monitoring and analyzing a baby's development, such as measuring handgrip strength, and safety considerations, such as preventing sudden infant death syndrome and detecting falls that may result in injuries.

Toddlers have difficulty communicating verbally, so VTAR is typically beneficial to their safety and well-being. The activities of adults have been extensively studied, but not the activities of toddlers. Adult humans will not perform many activities like toddlers for example crawling. In spite of the fact that toddlers perform many of the same tasks as humans, their manner of doing them is different. It is highly challenging to recognize toddler activities as opposed to human activities because toddlers' activities are highly subjective. Following are some instances.

- In contrast to an adult walking staircase, babies and toddlers may climb stairs by crawling. In addition, toddlers walk downstairs by crawling backward due to which it is impossible to determine whether the baby is going up or down the staircase from a single frame image.
- In comparison with adult sitting posture, a toddler sitting posture differs greatly. The various sitting postures include W-Sitting, Squat Sitting, Short Sitting, Tailor Sitting, Side Sitting, and Legs in front. W-sitting may lead to poor sitting balance, pigeon-toed walking, hip weakness, weak trunk and core muscles, poor trunk control, and tight hamstrings etc., [6, 7]. Therefore W-sitting postures in toddlers should be corrected and automated systems can be extremely helpful in this regard.
- A toddler's play is entirely different from a human's play. Toys are a favorite activity for kids. Identifying the different toys toddlers play with is essential to understand their behavior.
- Toddlers/Infants cries indicate emotional states, physical needs, and pathological problems caused by internal or external stimuli [8]. Toddlers cry for a variety of reasons, including hunger, pain, sleep, fear, or anxiety [9]. According to [10], a limited database resource hinders the development of infant cry research. It is imperative to have large databases with diverse samples suitable for research.
- It is more likely that toddler fall and be injured. As per the author [11] more than two million toddlers experience injury every year.
- Sleeping toddlers cry, smile, yawn, drink milk, and also move in bed more frequently, whereas adults do not do so.

Based on the above instances, authors conclude toddlers' activities are different from those of adults in terms of type and intensity, and therefore separate studies are needed. While toddler activity recognition is relatively unexplored, the next section summarizes some of the relevant works on infant and child activity recognition.

2 Related Works

Existing infant and toddler activity recognition systems leverage contact sensors. Devices with multiple sensors and other accessories are affixed to toddler skin with strong adhesives or tapes and are wired to large, bulky, and expensive base units (especially in the field of neonatal health care). Since toddler skin is very soft, all wearable sensors may cause skin irritation, and even toddlers tend to resist wearing sensors [12]. Furthermore, researchers in [13] state that a sensor's location affects its accuracy, so multiple sensors are used to achieve higher accuracy. Although sensor driven activity recognition has high performance, it is limited to basic activity recognition. According to the author in [5], activities like walking up and down stairs are difficult to classify accurately due to limitations in accelerometer signals. In order to detect more complex activities sensor-based systems must have a very complex architecture. This highlights the need for vision-based activity recognition systems. Since different camera types are available to provide more accurate data, vision methods are more advantageous than sensors.

Some notable research studies focused on the recognition of infant activity via vision have been published in [15–19]. Researchers in [15] propose a novel dataset to recognize infants' actions from videos in uncontrolled environments. It consists of 18 action classes including both normal and abnormal infant behavior. Up to the age of three years, 1181 video clips of infants were included in the dataset. However, the dataset is not publicly available. As a deep learning model, Convolutional Neural Networks (CNNs) are proposed to recognize child activities from videos with 94.73% accuracy. 155 videos from social media and the internet were used for the experiment (68905 frames in 155 videos) and the model was trained to detect four types of actions (crawling, running, sleeping, and walking) [16]. Using a Faster R-CNN and LSTM network, [17] proposed a multiview action classification system. Motion videos were recorded of infants (ages 8–24 months) in a constrained environment while the study involved four different action classes (crawl, sit, stand, and walk). A lightweight network, BabyNet (17 infant videos from YouTube), is proposed in [18]. This network is capable of modeling both short and long-range motion correlation related to different stages of reaching action. A smart baby monitoring system proposed by [19] detects baby objects using YOLO V7 and contains three object classes: baby, distressed baby, and fire. In this system, audio files are processed using Support Vector Machines (SVM) classifiers in order to identify mainly baby cries. Current activity recognition models face various limitations, including constraints related to the environment, the type of activity, the utilization of data acquisition devices, and the input data type (whether it is an image or a video). The subsequent section addresses the associated challenges with respect to VTAR.

3 Issues and Challenges

The HAR systems have evolved over the years, but state-of-the-art algorithms still pose several issues that result in erroneous classification of human activities or actions. The following are the issues and challenges related to VTAR.

- ***Dataset Limitations:*** Based on the survey papers [1, 2, 4, 14] on HAR, it was found that most of the existing HAR databases were only designed to detect adult actions due to a significant number of applications whereas toddlers have received very little research attention. Some benchmark databases include video evidence of babies in the adult dataset, for example, UCF50 (the "Swing class"), UCF101 (the "Baby Crawling and Swing class"), and HMDB-51 (the "Clap class") [20]. Currently, there is no publicly available dataset to analyze toddler activities. In order to carry out more research on young children, publicly available benchmark datasets are required. This is because the objective comparison between different machine learning approaches becomes possible with the increasing availability of publicly accessible datasets. Creating an efficient and benchmark dataset is highly challenging due to many inherent factors. Toddlers have fluctuating moods and they tend to change their behavior more frequently. Videotaping a toddler is a complex and delicate process that requires empathy, attention to detail, and adaptability. Therefore collecting large video clips of toddlers is difficult, time-consuming, and challenging [21].
- ***Activity Inference:*** Unlike the sensor-based approach, the vision-based approach does not determine toddler activity directly. It requires more complex processing to draw meaningful conclusions from images and videos. High-quality, representative data is needed for training. However, the number of videos in existing toddler action-specific datasets is insufficient to train deep-learning models for activity inference.
- ***Camera Field View:*** Devices in a VTAR are installed at fixed locations, which limits activity recognition to the camera viewing angle and makes it difficult to distinguish between two similar actions [22]. In order to solve this problem, CCTV cameras can be installed with auto-focus which intelligently adjusts the lens in order to focus on the object of interest [19]. The fact that toddlers move so often still makes it difficult.
- ***Occlusion:*** Occlusion is one of the biggest issues with respect to image and video processing. With respect to newborns and infants, self-occlusion is more common.
- **Inter-class Similarity:** In toddlers normal crawling and walking upstairs by crawling, walking, and running activities are very similar. Toddlers often crawl to climb staircases to differentiate between climbing staircases and normal crawling by toddlers the environmental context needs to be considered. Toddlers perform activities very quickly, resulting in video sequences that are typically of short duration. Additionally, many of these videos contain overlapping actions, which increases the similarity between different classes of activities.
- **Intra-class Variation:** The same toddler's pattern of performing an activity varies significantly compared to adults; for example, their ways of sitting, climbing stairs, walking, and crawling differ greatly.
- ***Computational Resources:*** Deep learning can solve HAR problems more accurately than machine learning algorithms [23]. Solving VTAR problem using deep neural networks requires substantial computational resources, including processing power, memory, and time. Therefore, conducting research is more expensive.
- ***Continuous Learning and Adaptability:*** It is challenging to enable VTAR models to adapt in real time and learn incrementally. "A combination of cameras and wearable sensors combined with deep learning can also improve the accuracy of activity recognition," states [22]. As a result, there is scope for integrated research on vision and sensor-based toddler activity recognition.

- ***Other Challenges:*** The other challenges for VTAR are the same as that of HAR, complex dynamic background, multi-view variation (front view and right view variations), illumination variation (light source and shadow), anthropometric variations, multi-view action fusion and recognition of more than one activity performed at the same time, etc. [1, 24–26].

4 Applications

VTAR holds potential applications in various domains such as video analysis, video transcription, video anticipation, toddler facial expression analysis, toddler monitoring systems, anomaly detection, infant motor development analysis, child-care training, and healthcare [12, 15, 23, 26].

5 Conclusion

In this research paper, computer vision-based technologies for the recognition of toddler activity is presented, with the authors finding that current studies on toddler activity recognition tend to identify and analyze specific action classes. Furthermore, there is no publicly available dataset for evaluating toddler activities. Due to their potential and practical applications in many emerging fields, the aforementioned problem is particularly interesting. In light of all the issues, challenges, and benefits mentioned above, VTAR has the potential for substantial growth in the future. Overall this study provides useful insight into the field of toddler activity recognition and further research in this direction.

References

1. Beddiar, D.R., Nini, B., Sabokrou, M., Hadid, A.: Vision-based human activity recognition: a survey. Multimed. Tools Appl. **79**(41–42), 30509–30555 (2020). https://doi.org/10.1007/s11042-020-09004-3
2. Zhang, S., Wei, Z., Nie, J., Huang, L., Wang, S., Li, Z.: A review on human activity recognition using vision-based method. J. Healthcare Eng. **2017**, 1–31 (2017). https://doi.org/10.1155/2017/3090343
3. Wikipedia contributors. Activity recognition. Wikipedia (2023). https://en.wikipedia.org/wiki/Activity_recognition
4. Girdhar, P.: Vision based human activity recognition: a comprehensive review of methods & techniques. Turkish J. Comput. Math. Educ. (TURCOMAT) **12**(10), 7383–7394 (2021). https://doi.org/10.17762/turcomat.v12i10.5631
5. Sindhu, P.: Toddler activity recognition using machine learning. Master's thesis. 3755 (2018). https://ecommons.luc.edu/luc_theses/3755
6. Admin. What is W-Sitting & How to Prevent it? | Child Development. Pathways.org (2022). https://pathways.org/what-is-w-sitting/
7. A Glossary of Sitting. Starfish Therapies (2010). https://starfishtherapies.wordpress.com/2009/04/27/a-glossary-of-sitting/
8. Ashwini, K., Vincent, P.M.D.R., Srinivasan, K., Chang, C.: Deep learning assisted neonatal cry classification via support vector machine models. Front. Public Health **9**, 670352 (2021). https://doi.org/10.3389/fpubh.2021.670352

9. De Bellefonds, C.: Five types of baby cries and what they might mean. BabyCenter (2023). https://www.babycenter.com/baby/crying-colic/different-baby-cries_40009946
10. Menin, D., Aureli, T., Dondi, M.: Two forms of yawning modulation in three months old infants during the face to face still face paradigm. PLoS ONE **17**(2), e0263510 (2022). https://doi.org/10.1371/journal.pone.0263510
11. Wang, P., Lien, S., Lee, M.: A learning-based prediction model for baby accidents. In: International Conference on Image Processing (2019). https://doi.org/10.1109/icip.2019.8803820
12. Hesse, N., Bodensteiner, C., Arens, M., Hofmann, U.G., Raphael Weinberger, A., Schroeder, S.: Computer vision for medical infant motion analysis: state of the art and RGB-D data set. In: Leal-Taixé, L., Roth, S. (eds.) ECCV 2018, pp. 32–49. Springer, Cham (2019). https://doi.org/10.1007/978-3-030-11024-6_3
13. Martínez-Villaseñor, L., Ponce, H.: A concise review on sensor signal acquisition and transformation applied to human activity recognition and human–robot interaction. Int. J. Distrib. Sens. Netw. **15**(6), 155014771985398 (2019). https://doi.org/10.1177/1550147719853987
14. Singh, R.K., Sonawane, A., Srivastava, R.: Recent evolution of modern datasets for human activity recognition: a deep survey. Multimed. Syst. **26**(2), 83–106 (2020). https://doi.org/10.1007/s00530-019-00635-732
15. Balasathiya, S.S., Roomi, S.M.M., Sathyabama, B.: Infant action database: a benchmark for infant action recognition in uncontrolled condition. J. Phys. **1917**(1), 012019 (2021). https://doi.org/10.1088/1742-6596/1917/1/012019
16. Suthar, B., Gadhiya, B.: Child activity recognition using deep learning. Int. J. Eng. Adv. Technol. **9**(5), 364–367 (2020). https://doi.org/10.35940/ijeat.E9563.069520
17. Pacheco, C., Mavroudi, E., Kokkoni, E., Tanner, H.G., Vidal, R.: A detection-based approach to multiview action classification in infants. In: International Conference on Pattern Recognition (2021). https://doi.org/10.1109/icpr48806.2021.9412822
18. Bhakri, V.: Infant Reaching Action Recognition in Unconstrained Environments (2021). https://escholarship.org/uc/item/81c6t7bt
19. Nandhakumar, R.G., Mohanapriya, S.: Smart baby monitoring system using YOLO V7 algorithm. In: 2022 International Conference on Information Technology Research and Innovation (ICITRI) (2022). https://doi.org/10.1109/icitri56423.2022.997
20. Huang, C., Wang, C., Wang, J.: Human action recognition system for elderly and children care using three stream ConvNet. In: International Conference on Orange Technologies (2015). https://doi.org/10.1109/icot.2015.7498476
21. Aljuaid, H., Mohamad, D.: Child video dataset tool to develop object tracking simulates babysitter vision robot. J. Comput. Sci. **10**(2), 296–304 (2014). https://doi.org/10.3844/jcssp.2014.296.304
22. Serpush, F., Menhaj, M.B., Masoumi, B., Karasfi, B.: Wearable sensor-based human activity recognition in the smart healthcare system. Comput. Intell. Neurosci. **2022**, 1–31 (2022). https://doi.org/10.1155/2022/1391906
23. Li, C., Pourtaherian, A., Van Onzenoort, L., Ten, W.E.T.A.: Infant facial expression analysis: towards a real-time video monitoring system using R-CNN and HMM. IEEE J. Biomed. Health Inform. **25**(5), 1429–1440 (2021). https://doi.org/10.1109/jbhi.2020.3037031
24. Manaf, A., Singh, S.: Computer vision-based survey on human activity recognition system, challenges and applications. In: International Conference on Signal Processing (2021). https://doi.org/10.1109/icspc51351.2021.9451736
25. Hosseini, A., et al.: Children activity recognition: challenges and strategies. In: International Conference of the IEEE Engineering in Medicine and Biology Society (2018). https://doi.org/10.1109/embc.2018.8513320

26. Efthymiou, N., Koutras, P., Filntisis, P.P., Potamianos, G., Maragos, P.: Multi-view fusion for action recognition in child-robot interaction. In: 2018 25th IEEE International Conference on Image Processing (ICIP) (2018). https://doi.org/10.1109/icip.2018.8451146

Automated W-Sitting Posture Detection in Toddlers

P. Shilpa(✉) and M. Chandrajit

Maharaja Research Foundation, Maharaja Institute of Technology Mysore, Mysuru, Karnataka, India

Shilpa628@gmail.com, chandrajithmmca@mitmysore.in

Abstract. In recent years, there has been an increasing demand for automated systems in various sectors, particularly in the healthcare industry. This research paper focuses on developing an automated system to address clinical issues related to toddler monitoring, specifically targeting the detection of W-sitting posture. While occasional W-sitting is considered typical in a toddler's development, habitual W-sitting can lead to atypical symptoms and potential health concerns. Therefore, it is crucial for parents to monitor their toddlers and discourage this sitting posture. This study used *RoboFlow 2.0 AutoML Classification model* to recognize W-sitting posture in toddlers automatically.

Keywords: Automated Systems · Computer Vision · Posture Detection · Toddler Monitoring · W-Sitting Posture

1 Introduction

Toddlers' sitting posture differs greatly from adult sitting posture [1]. W-Sitting is a posture where a child sits on the floor with their knees bent, legs splayed out to the sides, and their feet positioned outside of their hips. This posture is characterized by the formation of a "W" shape between the legs as shown in Fig. 1. On the other hand, normal sitting involves squat sitting, short sitting, side sitting, tailored sitting, and legs positioned in front, which offer better balance and stability. Potential risks and problems arising from habitual W-sitting include damage to the anterior cruciate ligament, back pain, poor sitting balance, delayed developmental milestones, pigeon-toed walking, hip weakness, and a swayback posture [1, 3, 4]. Improper sitting posture leading to back pain during childhood can pose a risk for lumbar disease in adulthood [5]. Therefore, cultivating correct postural habits from an early age is crucial. By monitoring a toddler's sitting habits and discouraging habitual W-sitting, parents and caregivers can promote better posture and movement patterns.

The International Hip Dysplasia Institute reports that W-sitting is often seen in toddlers around the age of 3, but fades as they grow [6]. The study [7] focused on investigating the impact of daily W-sitting on the standing posture of healthy toddlers aged 3 to 6 years. The study measured the pelvic tilt angle, which is the angle formed by

V. E. Balas et al. (Eds.): IEQIP 2023, LNNS 1029, pp. 157–163, 2024.
https://doi.org/10.1007/978-3-031-61929-8_23

a horizontal line and the line connecting two points of the iliac spine and the anterior superior iliac spine. Additionally, hip rotation was analyzed and classified into five groups using posture analysis techniques. The findings of the study revealed a correlation between W-sitting and a forward tilt of the pelvis, as well as internal rotation of the hip. This suggests that habitual W-sitting can lead to a standing posture characterized by hip internal rotation and forward pelvic tilt, even in toddlers with typical development. This research aims to use computer vision and machine learning models to distinguish between W-sitting and normal sitting in toddlers.

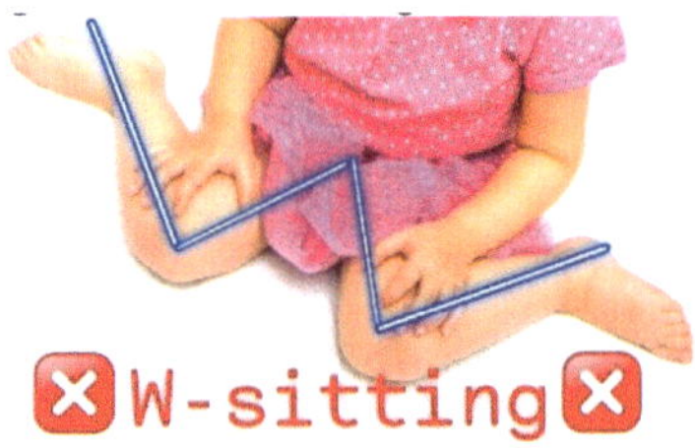

Fig. 1. W-Sitting Posture [2].

2 Related Works

Conventional methods for identifying W-Sitting patterns rely on manual observations by parents, healthcare professionals, and therapists. These assessments primarily depend on visual clues to differentiate between normal and W-sitting postures. Automating the classification process presents an opportunity to enhance objectivity and maintain consistent results. Notably, the application of computer vision to detect W-sitting postures has yet to be investigated. Within this context, the literature survey encompasses both children's and adult's sitting postures analysis.

Poor sitting posture can have long-term musculoskeletal consequences, particularly in children. To address this issue, a machine learning system utilizing pressure sensors, specifically a Convolutional Neural Network (CNN), was developed for classifying children's sitting postures with exceptional accuracy [8]. In the healthcare and assisted living sectors, a Deep Recurrent Hierarchical Network (DRHN) based on MobileNetV2 was introduced, with a focus on recognizing sitting postures [9]. Recognizing the significance of employee well-being in the workplace, especially concerning improper sitting postures, a smart chair system with IoT connectivity was created to detect and rectify incorrect postures, thereby improving both employee health and productivity [10]. Further emphasizing the health implications of prolonged incorrect sitting postures, a novel system called SPRS utilized machine learning and identified key chair surface areas to achieve an outstanding accuracy of up to 99.1% [11]. In educational settings, an innovative in-class sitting posture recognition system was introduced, employing OpenPose and a Keras-based CNN to detect and address improper student sitting postures, ultimately benefiting adolescents' well-being [12]. Additionally, [13] addressed health risks associated with poor sitting posture by developing a real-time posture monitoring system with an app interface.

In the healthcare sector, camera images are utilized to identify the initial posture of elderly patients, effectively preventing fall accidents as discussed in [14]. In the context of prolonged computer-related work and the associated risk of musculoskeletal disorders, a study successfully developed a detector for improper sitting postures using motion capture sensors, achieving an impressive accuracy of up to 98.4% [15]. For the aviation industry, where passenger comfort during flights is paramount, a study in [16] focused on detecting and classifying sitting postures using pressure data, achieving an 89.26% classification rate with support vector machines.

Innovations in posture detection extend to assistive technology as well, with anomaly detection for unbalanced sitting postures in wheelchairs using pressure sensor data, demonstrating promising results even without labeled data [17]. A real-time sitting posture detection system, grounded in deep learning and pressure sensors, was designed to correct poor sitting postures [18]. Models based on human skeleton key point detection and depth sensors achieved high classification accuracy [19, 20]. Non-contact detection using Frequency-modulated continuous wave radar (FMCW) reached high accuracy [21]. Additionally, motion sensors provided insights into functional fitness and well-being by detecting hand placements in elderly individuals when sitting and standing [22]. In [23] introduced a depth sensor-based sitting posture detection system that demonstrated superior recognition rates and speed compared to existing methods. Finally, models were developed to detect proper and improper sitting postures using gyroscope data from specific spinal points, emphasizing the importance of body frame characteristics in posture detection presented in [24].

The literature review uncovers a substantial body of research exploring diverse approaches to sitting posture analysis in both children and adults, encompassing sensor-based and vision-based methodologies. Nonetheless, there is a significant gap in existing research when it comes to the automated detection of W-sitting postures in toddlers. This research gap and the absence of relevant data serve as the primary driving force behind the current research initiative.

3 Proposed Method

Previous studies on the classification of sitting postures were not focused on W-sit-ting. The methodology employed in this research encompasses the following steps:

3.1 Dataset Description

To generate a dataset for model training, toddler images featuring recognizable W-sitting posture were obtained from prominent platforms like Google and through direct collection efforts. Due to the limited availability of images from Google search, a portion of the dataset was compiled through manual collection, including images sourced from personal efforts and selected frames from YouTube videos. Images are selected based on variations in environment and camera viewing angle for further processing. In total, 353 images were collected, of which 178 images depicted toddlers in W-sitting postures, while the remaining 175 images captured other sitting postures considered normal (all sitting postures except W-sitting). Consequently, our research focuses on a binary classification problem with "W-sitting" and "Normal sitting" as the two classes of interest.

The collected dataset is preprocessed to ensure uniformity and consistency. Image preprocessing, augmentation and model building are performed using RoboFlow online tools and services [25, 26]. Auto orient and resizing are the two image preprocessing techniques applied to the collected images. All images were resized to a uniform 320x320 resolution, with white edges padded to fit this size. To enhance model stability, image augmentation was employed, and 10% of the images were converted to grayscale, resulting in a total of 399 images. The dataset was then split into training, validation, and test sets, comprising 70%, 20%, and 10% images, respectively. The Fig. 2a and 2b depicts "W-sitting," and "Normal Sitting" postures in toddlers from the proposed dataset.

(a)

(b)

Fig. 2. (a) Sample images of W-Sitting in the proposed dataset. (b) Sample images of Normal Sitting in the proposed dataset.

4 Model Training and Evaluation

RoboFlow is an online computer vision platform that provides automated machine learning solutions for creating and training image classification models. The custom dataset was initially pre-trained using *RoboFlow 2.0 AutoML Classification model* with ImageNet as its backbone. The training loss and the validation accuracy graphs are in the Fig. 3a. The model's effectiveness was assessed using the validation set, achieving an accuracy of 87.8%. The validation set's confusion matrix is depicted in Fig. 3b. Precision, recall, accuracy, and F1 scores were calculated from the confusion matrix, represented by Eqs. (1), (2), (3), and (4) respectively.

$$Precision = \frac{TP}{TP + FP} = \frac{33}{33 + 6} = \frac{33}{39} = 0.846 \quad (1)$$

$$Recall = \frac{TP}{TP + FN} = \frac{33}{33 + 3} = \frac{33}{36} = 0.916 \quad (2)$$

$$Accuracy = \frac{TP + TN}{TP + TN + FP + FN} = \frac{33 + 32}{33 + 32 + 6 + 3} = \frac{65}{74} = 0.878 \quad (3)$$

$$F1Score = 2 * \frac{Precision * Recall}{Precision + Recall} = 2 * \frac{0.846 * 0.916}{0.846 + 0.916} = 2 * \frac{0.775}{1.762} = 0.879 \quad (4)$$

where TP is True Positive, FP is False Positive, FN is False Negative, and FP is False Positive.

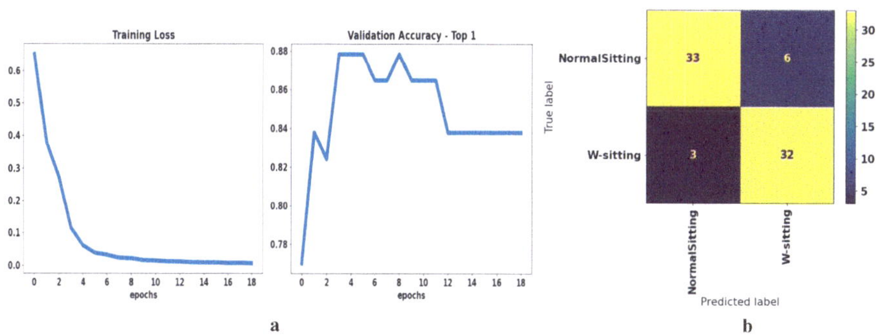

Fig. 3. (a) Training loss and validation accuracy graphs. (b) Confusion Matrix

The experimental results demonstrates good precision, recall, relatively high accuracy and the F1 score.

5 Conclusion and Future Scope

Occasional W-sitting in toddlers is commonly viewed as a normal behavior. However, persistent W-sitting raises concerns, prompting this study to identify W-sitting posture in images. The classification task was executed using *RoboFlow 2.0 AutoML Classification model*, which demonstrated a relatively high accuracy of 87.8%. Despite these positive outcomes, future research will focus on expanding the dataset and exploring alternative machine learning models to refine sitting posture classification. It is anticipated that incorporating a larger and more diverse dataset will enhance the reliability of the results.

References

1. W-sitting: Is it really a problem? (2023). https://www.healthline.com/health/baby/w-sitting
2. Don't let your baby sit like that! w-sitting dangers (2022). https://doi.org/10.1541/ieejeiss.139.1183. Accessed 27 Sept 2023
3. What to know about w-sitting in children (2023). https://www.webmd.com/children/what-to-know-about-w-sitting-in-children
4. What is w-sitting how to prevent it?—child development (2023). https://pathways.org/what-is-w-sitting/
5. Grimmer, K., Nyland, L., Milanese, S.: Longitudinal investigation of low back pain in Australian adolescents: a five-year study. **11**(3) (2006)
6. W-sitting and hip development - international hip dysplasia institute (2023). https://hipdysplasia.org/w-sitting-and-hip-development/
7. Fujimori, M., Natuzako, A., Tsurusaki, T.: Effect of w-sitting on standing posture of preschool age children. Rigakuryoho kagaku **30**, 429–432 (2015)
8. Kim, Y.M., Son, Y., Kim, W., Jin, B., Yun, M.H.: Classification of children's sitting postures using machine learning algorithms. **8**(8) (2018)
9. Kulikajevas, A., Maskeliunas, R., Damaševičius, R.: Detection of sitting posture using hierarchical image composition and deep learning. **7** (2021)
10. Matuska, S., Paralic, M., Hudec, R.: A smart system for sitting posture detection based on force sensors and mobile application. **2020** (2020)
11. Tsai, M.-C., Chu, E.T.-H., Lee, C.-R.: An automated sitting posture recognition system utilizing pressure sensors. Sensors **23**(13) (2023). https://www.mdpi.com/1424-8220/23/13/5894
12. Chen, K.: Sitting posture recognition based on openpose. **677**(3) (2019)
13. Laidi, R., Khelladi, L., Kessaissia, M., Ouandjli, L.: Bad sitting posture detection and alerting system using EMG sensors and machine learning. In: 2023 International Conference on Artificial Intelligence in Information and Communication (ICAIIC), pp. 324–329 (2023)
14. Inoue, M., Taguchi, R., Umezaki, T.: Sitting posture detection of patients pose estimation and bed detection. IEEJ Trans. Electron. Inf. Syst. **139**, 1183–1190 (2019)
15. Feradov, F., Markova, V., Ganchev, T.: Automated detection of improper sitting postures in computer users based on motion capture sensors. Computers **11**(7), 116 (2022). https://doi.org/10.3390/computers11070116
16. Cun, W., et al.: Sitting posture detection and recognition of aircraft passengers using machine learning. Artif. Intell. Eng. Des. Anal. Manuf. **35**(3), 284–294 (2021). https://doi.org/10.1017/S0890060421000135
17. Kini, K.R., Harrou, F., Madakyaru, M., Kadri, F., Sun, Y.: Automatic detection of unbalanced sitting postures in wheelchairs using unlabeled sensor data. IEEE Sens. Lett. **6**(8), 1–4 (2022)
18. Yang, P., Peng, Z.-L., Wang, W.-L.: Sitting posture detection system based on Keras framework. In: 2022 3rd International Conference on Computer Vision, Image and Deep Learning International Conference on Computer Engineering and Applications (CVIDL ICCEA), pp. 170–174 (2022)
19. Ji-sheng, F., Li, G., Xuan-xuan, L.: Sitting posture detection system based on stacking ensemble learning. In: 2022 IEEE 2nd International Conference on Computer Systems (ICCS), pp. 103–110 (2022)
20. Bei, S., Xing, Z., Taocheng, L., Qin, L.: Sitting posture detection using adaptively fused 3D features. In: 2017 IEEE 2nd Information Technology, Networking, Electronic and Automation Control Conference (ITNEC), pp. 1073–1077 (2017)
21. Liu, G., Li, X., Xu, C., Ma, L., Li, H.: FMCW radar-based human sitting posture detection. IEEE Access **11**, 102746–102756 (2023). https://doi.org/10.1109/ACCESS.2023.3312328

22. Muro, H., Shimakawa, H.: Detection of hand support at posture changes between sitting and standing in elderly daily living. Int. J. Intell. Inf. Process. **4**(2), 54–65 (2013). https://doi.org/10.4156/ijiip.vol4.issue2.6
23. Zeng, X., Sun, B., Luo, W., Liu, T., Lu, Q.: Sitting posture detection system based on depth sensor. Jisuanji Kexue = Computer Science **45**(7) (2018). Copyright Guojia Kexue Jishu Bu 2018. https://www.proquest.com/scholarly-journals/sitting-posturedetection-system-based-on-depth/docview/2249734201/se-2. Accessed 25 July 2019
24. Estrada, J.E., Vea, L.A.: Real-time human sitting posture detection using mobile devices. In: 2016 IEEE Region 10 Symposium (TENSYMP), pp. 140–144 (2016)
25. What is computer vision and machine vision? A guide for beginners (2023). https://blog.roboflow.com/intro-to-computer-vision/
26. September 2023. https://blog.roboflow.com/what-is-automl/

Enhancing Shear Wave Estimation Through Directional and Kalman Filters for Noise Reduction

Nguyen Sy Hiep[1], Tran Duc Nghia[2], Luong Quang Hai[3], Vu Thi Thuong[4], Kolla Bhanu Prakash[5], and Tran Duc Tan[6](✉)

[1] Thai Nguyen University of Information and Communication Technology, Thai Nguyen, Vietnam
Nshiep@ictu.edu.vn
[2] Institute of Information Technology, Vietnam Academy of Science and Technology, Hanoi, Vietnam
nghiatd@ioit.ac.vn
[3] Le Quy Don Technical University, Hanoi, Vietnam
luonghai@mta.edu.vn
[4] Phuong Dong University, Hanoi, Vietnam
thuongvt@dhpd.edu.vn
[5] K.L. Deemed to be University, Vaddeswaram, Andhra Pradesh, India
[6] Phenikaa University, Hanoi, Vietnam
tan.tranduc@phenikaa-uni.edu.vn

Abstract. In the present day, various imaging diagnostic techniques support doctors in the diagnosis of tumors, including MRI imaging, CT scans, X-rays, mode B ultrasound, and ultrasound elastography... Elastography techniques have garnered significant attention and strong development in recent years, offering advantages such as non-invasiveness, low cost, rapid examination times, and high reliability. To determine the elasticity of tissue, this technique requires estimating viscosity and elasticity based on the velocity of shear waves captured by the Doppler ultrasound probe. The velocity signal contains noise from reflected waves and random interference, affecting the signal quality. In this paper, the we used directional filter and Kalman filter to eliminated the noise.

Keywords: Shear Wave · Reflected Wave · Composite Wave · Directional Filter · Kalman Filter

1 Introduction

In this study we used a vibrating needle to generated shear waves. These shear waves propagated within the body's tissue. The healthy tissue has a different level of stiffness compared to diseased tissue. The speed of shear waves depends on the stiffness of the tissue [1]. We determine elasticity and viscosity from measured velocity Fig. 1.

However, the velocity signal is influenced by random noise and noise caused by reflected waves [1, 2]. There are many studies on elastography techniques. For example,

V. E. Balas et al. (Eds.): IEQIP 2023, LNNS 1029, pp. 164–169, 2024.
https://doi.org/10.1007/978-3-031-61929-8_24

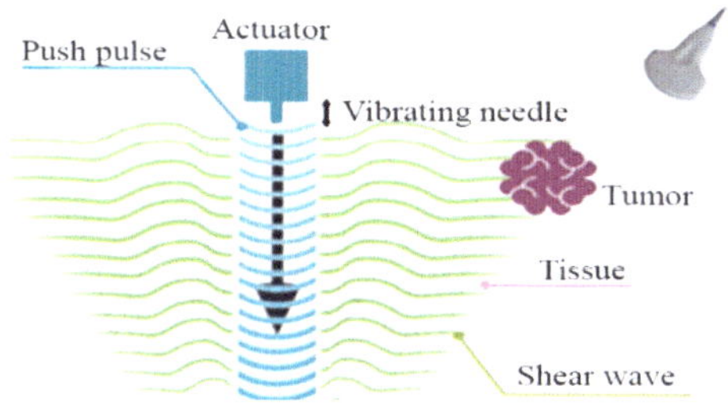

Fig. 1. Vibrating needle generate shear wave [1].

Yi Zheng used Kalman filters to eliminate random noise [3], while Thomas Defeux used directional filter to remove reflection noise [4]. In this study, we used the Kalman filter and the directional filter to eliminated both types of noise mentioned above.

2 Materials and Methods

2.1 The Stiffness of Living Tissues

$$E = \frac{s}{e} \tag{1}$$

where E is Young Modulus, s is the external stress, e is the strain inside the tissue. Softer tissue has a lower density than hard tissue. The Young's modulus parameter is used to determine the stiffness of tissue in the human body. The elasticity of healthy breast tissue is within the range of 18 to 24 kPa, while in breast tumor tissue, it ranges from 22 to 560 kPa. by the same token that diseased tissue shows greater stiffness compared to healthy tissue. Young modulus parameter is calculated using the following formula [1, 5, 6].

$$E = \rho c^2 \tag{2}$$

where ρ is density of tissues (1000 kg/m^3), c is Shear wave propagation speed is external stress, e is strain inside the tissue, E is Young Modulus. Shear waves propagate at a speed of 1 to 10 m/s [5, 6].

2.2 Estimating the Elasticity and Viscosity of Tissue.

To determine the elasticity and viscosity of tissue in this study, we employed wave equations to simulated the propagation of shear waves in the tissue [6].

The equation of the incoming wave (3).

$$v_n(r) = \frac{1}{\sqrt{r - r_0}} A e^{-a(r-r_0)} \cos[\omega n \Delta t - k_s(r - r_0) - \phi] \tag{3}$$

When a deformation waves reach the boundary between two tissue environments with different density ratios, it produces a reflected wave. The equation for the reflected wave (4)

$$v_n(r) = \frac{1}{\left(\sqrt{r_{f0} - r_f}\right)} \frac{A}{C} e^{-a(r-r_0)} \cos[-\omega n \Delta t + k_s(r_{f0} - r_f) - \phi] \quad (4)$$

In the Kelvin-Voigt model, determining the elasticity (E) is synonymous with calculating the complex shear modulus μ (5) [2, 7].

$$\mu = \mu_1 - i\omega\eta \quad (5)$$

$$\mu_1 = \frac{\rho\omega^2(k_s^2 - \alpha^2)}{(k_s^2 + \alpha^2)^2} \quad (6)$$

$$\eta = -\frac{2\rho\omega k_s \alpha}{(k_s^2 + \alpha^2)^2} \quad (7)$$

The determination of elasticity μ_1 and viscosity η relies on Eqs. (6) and (7).

3 Result and Discussion

3.1 Data Processing Flow

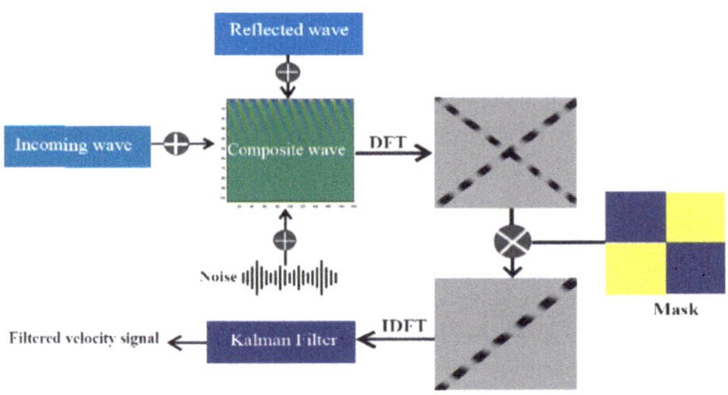

Fig. 2. Flow of Data Processing.

As mentioned earlier, the estimated velocity signal contains various types of noise, including reflected noise and random noise. This data underwent Fourier transformation before being multiplied by a directional filter matrix to remove reflected waves Fig. 4. Subsequently, the signal passed through a Kalman filter to eliminate random noise (Fig. 2). The velocity signal, after being filtered, was used to estimate the elasticity and viscosity of the living tissues [8–10].

In this paper, we utilized MATLAB for simulation purposes, and the input data was obtained from the values provided in Table 1.

Table 1. The elasticity and viscosity of the medium

Spatial locations	Elasticity (Pa)	Viscosity (Pa.s)
1–50	650	0.1
51–105	800	0.2

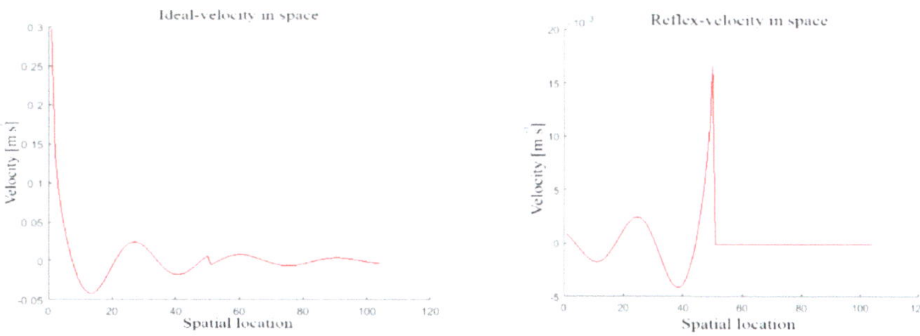

Fig. 3. (a) The incoming wave (b) The reflected wave.

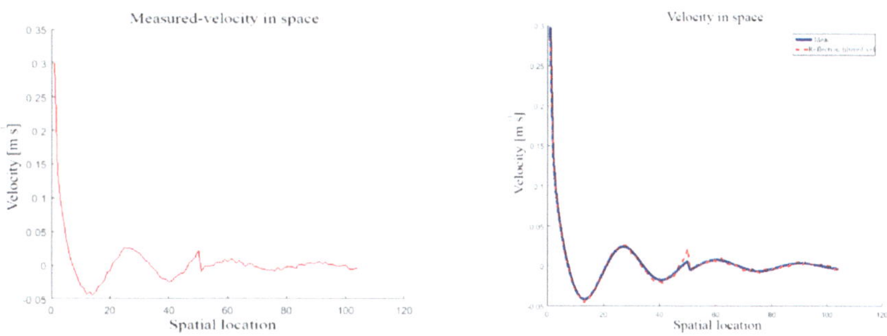

Fig. 4. (a) The composite wave (b) The composite after using filters

3.2 Result

We evaluated the filters' influence by comparing the results both before and after applying filters.

The velocity signal obtained (Fig. 4a) includes reflected (Fig. 3b) and random noise. Observing the graph Fig. 4a. It is evident that the impact of noise is significant, especially at position 50. The velocity signal after undergoing filtering, shows reduced noise in the dashed line graph in Fig. 4b. This signal closely matches the ideal graph represented by the blue line (Fig. 5).

The velocity graph at position 25 represents a waveform with a constant amplitude. To evaluate the effectiveness of the applied filter, in figure (a), the measured velocity graph before filtering is compared with the ideal velocity graph, while in figure (b), the

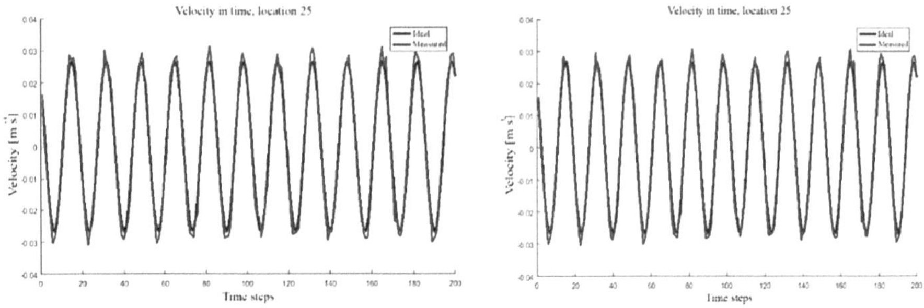

Fig. 5. Velocity graph at position 25. (a) Before using the filters. (b) After using the filters.

signal after filtering is shown. It can be observed that the signal in figure (b) closely matches the ideal graph compared to figure (a) (Fig. 6).

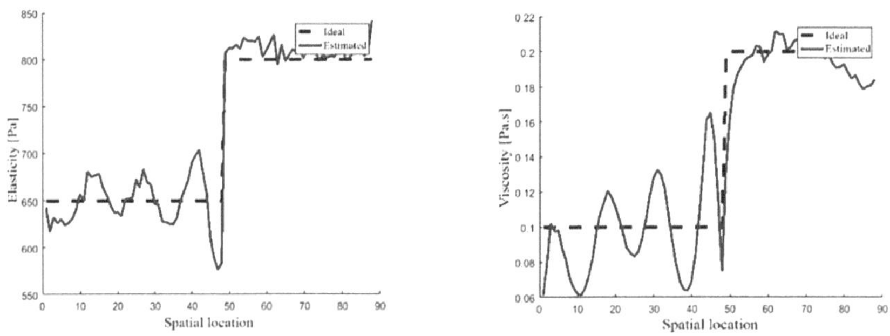

Fig. 6. (a) Elasticity graph (b) Viscosity graph

The velocity signal, after being filtered, has been used to estimate elasticity and viscosity in Figures A and B. The dashed line graph represents the ideal graph, and it can be noted that the estimated signal closely tracks the values of the ideal graph. Both the ideal and estimated graphs show a strong response at position 50. The values on the graphs align well with the values in Table 1.

4 Conclusions

By employing directional filter and Kalman filter, we have successfully reduced the noise affecting signal quality. As a result, we have achieved more reliable estimates of elasticity and viscosity. This has significant implications for diagnostics. In the future, we aim to further develop this technique to optimize signal quality, potentially even constructing 2D or 3D images of tissue elasticity and viscosity.

Funding. This work was supported by National Foundation for Science and Technology Development (NAFOSTED) under Grant 103.05-2020.13.

References

1. Hiep, N.S., Hai, L.Q., Nghia, T.D., Tan, T.D.: Evaluating the improvement in shear wave speed estimation affected by reflections in tissue. In: Nguyen, T.D.L., Verdú, E., Le, A.N., Ganzha, M. (eds.) ICISN 2023, vol. 752, pp. 441–451. Springer, Singapore (2023). https://doi.org/10.1007/978-981-99-4725-6_54
2. Song, S., Le, N.M., Huang, Z., Shen, T., Wang, R.K.: Quantitative shear-wave optical coherence elastography with a programmable phased array ultrasound as the wave source. Opt. Lett. **40**(21), 5007 (2015). https://doi.org/10.1364/OL.40.005007
3. Zheng, Y., Chen, S., Tan, W., Kinnick, R., Greenleaf, J.F.: Detection of tissue harmonic motion induced by ultrasonic radiation force using pulse-echo ultrasound and Kalman filter. IEEE Trans. Ultrasonics Ferroelectrics Frequency Control **54**(2), 290–300 (2007). https://doi.org/10.1109/TUFFC.2007.243
4. Deffieux, T., Gennisson, J.-L., Bercoff, J., Tanter, M.: On the effects of reflected waves in transient shear wave elastography. IEEE Trans. Ultrasonics Ferroelectrics Frequency Control **58**(10), 2032–2035 (2011). https://doi.org/10.1109/TUFFC.2011.2052
5. Luong, Q.-H., Nguyen, M.-C., Tran-Duc, T.: Complex shear modulus estimation using extended Kalman filter. Tạp chí khoa học& kỹ thuật số 179 (2016)
6. Carlson, L.C., Feltovich, H., Palmeri, M.L., Dahl, J.J., Munoz del Rio, A., Hall, T.J.: Estimation of shear wave speed in the human uterine cervix. Ultrasound Obstet. Gynecol. **43**, 452–458 (2014)
7. Yuan, H., Guzina, B.B., Chen, S., Kinnick, R.R., Fatemi, M.: Estimation of the complex shear modulus in tissue-mimicking materials from optical vibrometry measurements. Inverse Probl. Sci. Eng. **20**(2), 173–187 (2011). https://doi.org/10.1080/17415977.2011.589903
8. Huy, T.Q., Ha, P.T.T., Duong, T.B., Vinh, N.Q., Yen, N.T.H., Tan, T.D.: 2D shear wave imaging in Gaussian noise and reflection media. VNU J. Sci. Math. Phys. **37**(4) (2021). https://doi.org/10.25073/2588-1124/vnumap.4631
9. Hao, N.T., Thuy-Nga, T., Dinh-Long, V., Duc-Tan, T., Linh-Trung, N.: 2D shear wave imaging using maximum likelihood ensemble filter. In: International Conference on Green and Human Information Technology (ICGHIT 2013) (2013)
10. Song, P., Manduca, A., Zhao, H., Urban, M.W., Greenleaf, J.F., Chen, S.: Fast shear compounding using robust two-dimensional shear wave speed calculation and multi-directional filtering. Ultrasound Med. Biol. (2014)

Developing Efficient Methods for Tracking the Location, Status, and Maintenance Needs of Oxygen Cylinders Equipped with IoT Devices

Chinthakuntla Vignesh[1], R. Jagadesh[1(✉)], K. Ashifa[1], N. R. Subesh[1], and T. Kiruba Devi[2]

[1] Department of Computer Science and Engineering, Dr. M.G.R Educational and Research Institute, Chennai, Tamilnadu, India
jagadeshradha27@gmail.com

[2] Department of Computer Science and Engineering (Artificial Intelligence), Dr. M.G.R Educational and Research Institute, Chennai, Tamilnadu, India

Abstract. In order to effectively manage oxygen cylinders in healthcare facilities, this paper investigates Internet of Things-based strategies. It discusses tracking, security, and predictive maintenance issues while highlighting how IoT can enhance patient care and inventory management. Data collection, sensor tagging, and device deployment are described in the study, with a focus on hospital infrastructure integration and regulatory compliance. Staff satisfaction, location accuracy, and successful predictive maintenance are all indicated by the results. Notwithstanding certain drawbacks, the paper promotes broad IoT adoption in healthcare facilities. This contribution broadens our understanding of healthcare technology and offers potential directions for future IoT development in resource management.

Keywords: Oxygen cylinder management · Healthcare facilities · Internet of Things (IoT) · Sensor technology · Predictive maintenance · Inventory management · Security

1 Introduction

This study addresses the imperative need for precise oxygen cylinder tracking in healthcare, aiming to enhance patient safety and streamline medical care. Traditional manual tracking methods have proven error-prone, leading to challenges like unknown cylinder locations and delayed maintenance detection. Introducing an Internet of Things (IoT)-based system, this work focuses on achieving accurate real-time tracking, predictive maintenance, and adherence to healthcare regulations. The study outlines objectives, including IoT system implementation, precision evaluation, predictive maintenance algorithm application, and strict adherence to security and healthcare guidelines. By leveraging IoT devices with sensors and connectivity, this approach automates crucial aspects of healthcare resource management, optimizing productivity and patient care

V. E. Balas et al. (Eds.): IEQIP 2023, LNNS 1029, pp. 170–175, 2024.
https://doi.org/10.1007/978-3-031-61929-8_25

[1]. The paper provides a comprehensive analysis of IoT-driven strategies for monitoring oxygen cylinders, emphasizing their significance in improving healthcare operations and patient outcomes.

2 Related Work

The provided literature comprehensively explores the impact of Internet of Things (IoT) technology on healthcare, with a specific focus on oxygen cylinder tracking. Authors such as Bard, Ahmed, and Khan emphasize IoT's practical applications in predictive maintenance, real-time tracking, and operational acceleration, ensuring a reliable oxygen supply and enhancing patient safety [1]. Kumar et al. (2022) present an IoT-based system for healthcare, tracking oxygen cylinders through GPS, pressure sensors, and RFID labels. Emphasizing enhanced efficiency, the study aims to improve healthcare management by enabling real-time monitoring and predictive maintenance alerts. In contrast, Farooqui et al. (2022) explore an IoT-based automated greenhouse using machine learning, aiming to optimize conditions and boost productivity in greenhouse operations [2]. Ali's work underscores the importance of data privacy and legal compliance in deploying IoT solutions within healthcare organizations. Wang, Lee, and Chen exemplify the adaptability of IoT solutions in diverse healthcare contexts, emphasizing customization for specific needs like data security and asset tracking. Liu and Kumar's research anticipates future developments in IoT-enabled healthcare, emphasizing the need for continual adaptation to innovations in this dynamic sector [3]. The study concludes with insights into the integration of the IoT system with hospital infrastructure, presenting results on accuracy, employee satisfaction, and system scalability.

3 Overview of Oxygen Cylinder System

Hospital oxygen cylinder systems are intended to provide medical-grade oxygen to patients requiring oxygen therapy for a range of medical conditions, as well as patients undergoing surgery or respiratory disorders. To guarantee the efficient and safe delivery of oxygen, it is comprised of a well-organized infrastructure that includes oxygen cylinders, a centralized distribution system, pressure regulators, flow meters, and an alarm system [2]. The oxygen cylinders are the central component of the oxygen cylinder system. These cylinders, which can hold compressed oxygen suitable for medical use, are usually constructed from high-strength materials like steel or aluminum. Depending on the particular requirements of the healthcare facility, they are available in a range of sizes and capacities, from small portable cylinders to large stationary ones [3].

In hospital settings, a centralized oxygen storage area, equipped with ultrasonic concentration sensing, adheres to safety standards. Pressure regulators ensure oxygen delivery at a safe and consistent pressure from each cylinder. Flowmeters, with a liters-per-minute scale, allow precise adjustment for patient needs. Piping connects storage to patient areas, ensuring efficient oxygen distribution. Backup systems, including cylinders and concentrators, maintain supply during emergencies. Monitoring and alarm systems detect issues, alerting healthcare staff. Patient care involves constant monitoring to ensure equipment efficacy. System implementation integrates the IoT system with hospital infrastructure, yielding accurate and dependable results. The study discusses limitations, interpret results, and emphasizes system adaptability and scalability in the discussion section (Fig. 1).

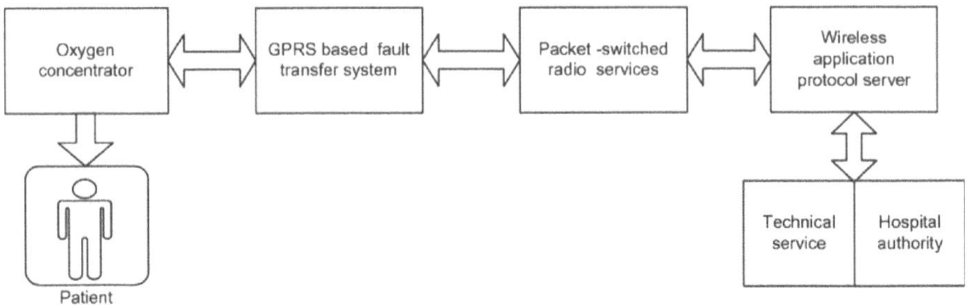

Fig. 1. Process diagram of oxygen distribution network in a healthcare division

4 Existing System

Patient care and safety may be jeopardized by the problems in healthcare settings associated with poor inventory control, misplaced cylinders, and neglected maintenance of oxygen cylinders. Systems for manually tracking inventories can be labor-intensive and prone to errors, which can result in inefficient use of resources and serious problems like oxygen shortages during medical treatments. In addition, the constantly changing environment of hospitals, where oxygen cylinders are routinely moved, might make emergency preparedness and prompt crisis response more difficult [6]. Additional dangers arise from neglecting safety compliance and maintenance. Cylinders may be used past their safe operating life or without required inspections, which would violate healthcare laws and jeopardize patient safety.

All of these issues are addressed in full by the proposed Internet of Things (IoT)-based oxygen cylinder tracking system. It provides accurate inventory control, improves emergency readiness, and encourages compliance with safety requirements by using real-time tracking and predictive maintenance algorithms. This technology innovation solves the main issues related to manual inventory management, cylinder misplacement, and careless maintenance in healthcare settings, while also streamlining hospital operations and protecting patient care

5 Proposed System

The proposed solution employs a comprehensive system integrating IoT-equipped oxygen cylinder sensors. Each cylinder features GPS for real-time location, enhancing emergency response. Pressure sensors monitor oxygen levels, predicting refill needs. Temperature and humidity sensors ensure safe storage. RFID labels provide unique IDs for efficient inventory management. IoT sensors wirelessly transmit data to a central database, including location, pressure, temperature, and maintenance logs. Historical data enables analytics, trend analysis, and predictive maintenance using machine learning algorithms. Automated alerts and maintenance schedules ensure timely interventions. Hospital staff access real-time data and analytics through online and mobile applications, improving operational efficiency.

- Check the location of oxygen cylinders
- Receive maintenance alerts and schedules
- View historical data and trends
- Monitor the overall status of the oxygen supply (Fig. 2)

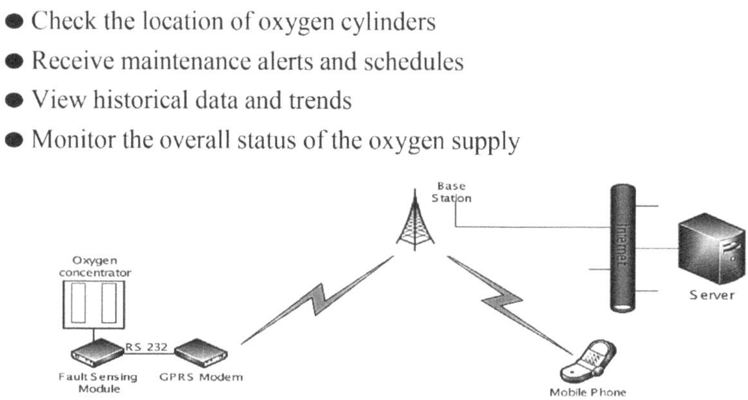

Fig. 2. Oxygen monitoring system

A well-defined process that includes system setup, data collection, storage, and predictive maintenance is involved in the installation of an Internet of Things-based oxygen cylinder monitoring system in a medical environment. The following steps can be used to describe how the system operates: The first step in the process is to choose and set up the right IoT devices. A central server receives the data gathered from the Internet of Things devices and uses it for real-time monitoring and analysis. The central element of the system is this server, which gathers information from each cylinder located throughout the medical centre. Because the server runs on a cloud- based platform, authorised staff can monitor it remotely and access data from any location within the building.

5.1 Visualisation

Healthcare personnel can view and manipulate the data produced by IoT devices using an intuitive interface. The purpose of this interface is to provide up-to-date information on the position, condition, and required maintenance of every oxygen cylinder. Visual aids, like interactive maps or dashboards, make it possible for employees to locate cylinders precisely inside the building with ease. Graphs and charts are used to display historical data and usage patterns in addition to real-time tracking [9]. This historical data visualisation helps with trend identification, inventory management optimisation, and data-driven cylinder maintenance and replacement decisions

5.2 Techniques for Integration

The ThingSpeak server and the oxygen monitoring system are integrated using tried- and-true methods for data management and IoT connectivity. The methods listed below are used to ensure seamless integration.Data Transmission Protocols: IoT devices securely send data (Fig. 3) to the ThingSpeak server using common data transmission protocols like HTTP and MQTT (Message Queuing Telemetry Transport). Data transfer is dependable and effective thanks to these protocols.

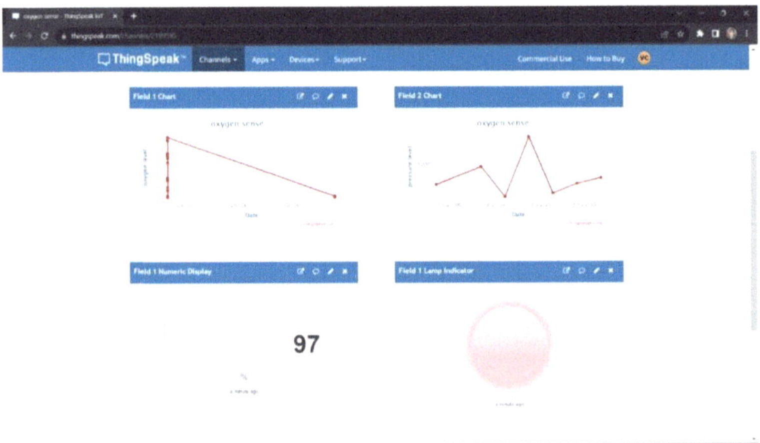

Fig. 3. Visualization of generated data in Thingspeak server

IoT data integrates through ThingSpeak API, enabling tailored healthcare configurations. A scalable cloud database accommodates expanding oxygen needs, with alerts for swift responses to issues. Integration with the hospital's Health Information System ensures seamless operations. Safety checks monitor temperature and pressure, triggering alerts. In emergencies, the system prioritizes oxygen distribution, enhancing safety and response. This method guarantees a reliable oxygen supply, elevating patient care and hospital efficiency.

6 Conclusion

In conclusion, the implementation of an Internet of Things (IoT)-based oxygen cylinder monitoring system in medical institutions is a huge step forward, addressing difficulties such as manual inventory management, cylinder misplacement, and maintenance neglect. This system makes use of Internet of Things technology to enable effective resource allocation, predictive maintenance, and real-time tracking, allowing for precise cylinder handling, adherence to safety rules, and improved inventory control, eventually assuring a steady oxygen supply critical for patient care. Integration with the cloud-based ThingSpeak platform makes data available for informed decision-making and proactive response, establishing a precedent for IoT adoption in healthcare, providing long-term adaptability and scalability, and revolutionizing healthcare technology and resource management for better patient care in medical settings.

References

1. Kumar, A., Singh, P., Kumar, V.: IoT-based oxygen cylinder tracking system for enhanced healthcare management. J. Med. Syst. **46**(12), 1–10 (2022)
2. Farooqui, N.A., Mishra, A.K., Mehra, R.: IOT based automated greenhouse using machine learning approach. Int. J. Intell. Syst. Appl. Eng. **10**(2), 226–231 (2022)
3. Farooqui, N.A., Ritika: A machine learning approach to simulating farmers' crop choices for drought prone areas. In: Singh, P., Panigrahi, B., Suryadevara, N., Sharma, S., Singh, A. (eds.) Proceedings of ICETIT 2019. LNEE, vol. 605, pp. 472–481. Springer, Cham (2020). https://doi.org/10.1007/978-3-030-30577-2_41
4. Khan, M., Ahmad, S., Khan, J.: A Novel IoT-enabled approach for oxygen cylinder tracking and management in hospitals. Sens. Actuat. A **348**, 114126 (2022)
5. Li, W., Zhang, H., Zhang, J.: IoT-driven oxygen cylinder tracking and management system for enhanced healthcare efficiency and patient safety. IEEE J. Biomed. Health Inform. **26**(7), 1465–1475 (2022)
6. Chen, T., Wang, M., Chen, Z.: IoT-based oxygen cylinder tracking and monitoring system for improved efficiency and safety in healthcare. Int. J. Med. Inform. **167**, 104962 (2022)
7. Liu, Y., Huang, J., Liu, G.: Edge computing-enabled IoT-based oxygen cylinder tracking system for real-time monitoring and predictive maintenance in healthcare. J. Healthc. Eng. **23**(5), 508–522 (2022)
8. Kim, D., Kim, E., Kim, S.: AI-powered IoT-based oxygen cylinder tracking and predictive maintenance system for healthcare facilities. JMIR Med. Inform. **25**(4), e41202 (2023)
9. Smith, A., Jones, B., Smith, C.: IoT-based oxygen cylinder tracking system for healthcare facilities: a comprehensive review. Digital Health **9**(2), 20552005 (2023)
10. Garcia, W., Lopez, H., Garcia, J.: IoT-driven oxygen cylinder tracking and management system for enhanced healthcare efficiency and patient safety. IEEE Trans. Biomed. Eng. **69**(6), 2315–2325 (2022)

Classification of Cyber Attacks in Internet of Medical Things Using Particle Swarm Optimization with Support Vector Machine

J. Viji Gripsy[1](✉), M. Sasikala[1], and R. Maneendhar[2]

[1] PSGR Krishnammal College for Women, Coimbatore, Tamilnadu, India
vijigripsy@gmail.com, sasikalam@psgrkcw.ac.in
[2] KG College of Arts and Science, Coimbatore, Tamilnadu, India
r.maneendhar@kgcas.com

Abstract. The Internet of Medical Things (IoMT) is a tangible application of the IoT inside the healthcare sector, presenting an opportunity to enhance the existing healthcare system and enhance patient well-being and quality of life. The presence of vulnerabilities in the ongoing transmission of IoMT data is a potential risk, since it enables unauthorised individuals to criminally get access to sensitive health information. This compromised data might then be used for malicious purposes. A multitude of security approaches have been suggested and developed in order to effectively mitigate the security problems associated with the IoMT. This paper aims to frame cyber-attack prediction as a classification issue, with a focus on the field of networking. The objective is to predict the specific kind of network attacks using machine learning algorithms. The classification of different forms of Denial of Service (DOS) assaults includes the following: BENIGN, MSSQL, DNS, LDAP, TFTP, SSDP, SYN Flood, PORTMAP, SNMP, NetBIOS, UDP, UDP-Lag, and WebDDoS. The findings indicate that the suggested PSO with SVM algorithm has a 97% accuracy rate, outperforming than existing algorithms.

Keywords: Internet of Medical Things · Cybersecurity · Machine Learning · Optimization Healthcare Analytics

1 Introduction

Cybersecurity is crucial as healthcare institutions employ more digital technologies and store sensitive patient data electronically. Cyberattacks target vital healthcare data [11]. Financial, PHI, and other sensitive data may be exploited. Healthcare facilities need encryption, safe data storage, software upgrades, staff training, and incident response to decrease these hazards [12]. Health data hacks may compromise sensitive data, damage healthcare institutions, and cost money. Health data hackers steal identities, commit fraud, and more [13]. Healthcare facilities with large financial data sets are hackable. Healthcare organizations should utilize encryption, safe data storage, software upgrades, staff training, and incident response to prevent attacks [14]. Avoid disclosing important health information online, use safe passwords, and monitor bank accounts for odd activity [15].

V. E. Balas et al. (Eds.): IEQIP 2023, LNNS 1029, pp. 176–183, 2024.
https://doi.org/10.1007/978-3-031-61929-8_26

Cyberattacks on health data may steal PHI, harm healthcare companies' reputations, and cost money. Cybercriminals want health data for identity theft, fraud, etc. [6]. Several hospitals save financial data, making them hackable. Healthcare organizations should utilize encryption, secure data storage, software upgrades, staff training, and well-defined incident response techniques to prevent these attacks. To protect health information, consumers should use strong passwords, avoid disclosing personal information online, and check bank accounts for odd activity [7]. Most healthcare organizations worry about IoMT and IoT device security. Increase IT asset visibility and OT system security. Threat mitigation reduces security events' likelihood and impact. Strong passwords, data encryption, regular security audits, and software/firmware upgrades fix problems. Organizations require robust incident response systems to quickly fix security vulnerabilities and limit their impact [8]. IoMT includes smart drugs, telemedicine, and internet-connected wearable health monitoring. These gadgets [17] may provide physicians health data for remote monitoring and therapy [9]. Online health data storage on cloud infrastructures by IoMT devices increases security risks. Reduce these vulnerabilities by prioritizing IoMT device and data security. Safe program design, security upgrades, encryption, and secure data storage and transport protocols can do this. Healthcare firms must regularly assess IoMT system security and develop breach response plans [10].

2 Literature Review

AdaBoost, the premier DDoS detector, uses CICIDS2017 [9]. The model has 0.77 F1, 0.84 recall, and 0.77 accuracy. SMOTE enhanced dataset minority class values and model sensitivity. PCA added 16 dataset characteristics after 25 ensemble selection. Their dataset showed AdaBoost 0.81 accuracy. Kurniabudi et al. [4] identified three weighted CICIDS2017 subgroups using information gain feature selection. Twenty percent was training, thirty testing. Testing data was 30% and training 70%. Every group received one of five classifiers. With 99.86% accuracy, the 22-feature RF classifier model performed best. Additionally, the 52-feature J48 classifier model was 99.87% accurate. RF processing is faster than J48 classifier processing.

Abualkibash and Pelletier used R's Boruta package to evaluate CICIDS2017 [16]. Machine learning and ANN categorized RF data. The average RF model test data attack type identification accuracy over 68.35 h was 96.24%. Assault situations make the 500-repetitive ANN less consistent than the RF model. Aamir et al. [5] examined Friday afternoon CICIDS2017 data. The data includes DDOS, Benign, and PortScan. Twelve features—including infinite and monovalent values were lost. Features removed had <20% attribute correlations. Standard scalar normalisation was applied to 21 new data characteristics. The dataset was 70% instructive, 30% experimental. This research employs decision trees, discriminant analysis, SVM, nearest neighbor, and community classifiers. Classifier accuracy: 60.6%, 97.1%, 99.0%, 68.7%, 85.5%.

Yi and Aye (2019) created a CICIDS2017 DoS/PortScan detection model. The model has 6 classifiers. RF detects DOSattacks 99.799%, whereas SVM and JriP 100% identify PortScan attacks. A deep learning model was trained using CICIDS2017 and NSL-KDD [1]. Classifier accuracy is 99.43% for CICIDS2017 and 99.63% for NSL-KDD.

Previous studies were less accurate than DNN. To simulate enigmatic, ambiguous, or sophisticated intermediate processes, prediction analysis employs several methods and technologies. Data from the past and present informs predictions. An early warning system's network offense and defense is DoS prediction. DoS assaults are reduced via abnormality detection. Research explored DoS assaults from several angles. However, these methods needed previous information and rendered DoS assaults hard to identify from burst traffic. Predicting these assaults needs historical data.

3 Methodology

3.1 Proposed PSO with SVM

The proposed SVM is used for the development of a predictive model aimed at diagnosing and predicting diseases within the healthcare sector. The use of Particle Swarm Optimization (PSO) was employed in order to determine optimal support vectors that would effectively boost classification accuracy by increasing the margin.

Usage of Particle Swarm Optimization

PSO is a popular global optimization method used in many engineering fields. The velocity parameter in the PSO method helps particles travel across the search space. PSO uses particle velocity to identify direction. The technique modifies particle direction iteratively. The PSO method links each particle to three location-dependent parameters. These three criteria are important: 1) current location, 2) previous highest position, and 3) velocity.

Two positions are encountered while computing the particle's location. Optimization issues have global and local optimal solutions. Local best position gathers information from neighboring particles, whereas global best position gathers information from all particles in the system.

Parameter Selection

Let search space consist of d-dimensions during the nth iteration of the "i" swarm particle and is represented in Eq. (1)

$$Xi(n) = \{xi1(n), xi2(n), xi3(n), \ldots .xid(n)\} \tag{1}$$

The previous best swarm particle position is considered as $X^{previous-best}{}_i$, (n), and the next best swarm particle is described as $x^{next-best}{}_i(n)$.

The new particle position "i" is considered in Eq. (2)

$$X_i(n+1) = X_i(n) + v_i(n+1).\Delta t \tag{2}$$

Here vi describes the velocity of the particle concerning time. At represents the change in time. The velocity of the particle is associated with weights such as interia, cognitive and social and represented as ω, cl and ci respectively. It is represented in the Eq. (3)

$$v_i(n+1) = \omega \cdot v_i(n) + c_1 \cdot D_1(n) \cdot \left(x_i^{previous-best}(n) - x_i(n)\right)$$

$$+ c_2 \cdot D_2(n) \cdot \left(x_i^{nextbest}(n) - x_i(n)\right) \quad (3)$$

Interia is the velocity that provides information about the continuity of the particle velocity from iteration to iteration. Cognitive velocity is represented to move the particle to its previous best position. Social in the velocity represented to move the best particle towards its neighbourhood.

Implementation of Proposed SVM

In SVM, a hyperplane separates the dataset into two different classes. The performance of the SVM depends on the hyper plan. SVM uses a kernel function to improve performance. This kernel function has parameter "γ" and regularization parameter "C" to improve classification accuracy. Here PSO is used to identify parameter value "γ" and regularization parameter value.

Equation (4) represents the radial bias-based kernel function for the SVM

$$K(x_1, x_2) = \exp(-\gamma \|x_1 - x_2\|^2) \quad (4)$$

Here x_1, x_2 are considered data elements, Euclidian distance is represented as $\|x_1 - x_2\|$ and "γ" describes the decision region.

Margin "M" is directly associated with misclassification rate "MCR" and is represented in the Eq. (5)

$$MCR \, \alpha \, M \quad (5)$$

Margin "M" is indirectly associated with regularization parameter "C" and is represented in Eq. (6)

$$M \, \alpha \, \frac{1}{c} \quad (6)$$

The training model takes the input data "X" with the class label "γ". The learning rate in the SVM specifies choosing of parameters.

Procedure for Using PSO in SVM

The optimization process starts from selecting the parameters ω, cl, c2 assigning the values for these parameters and then position and velocity of the particle is defined. Then SVM model was trained with SVM parameters "γ" and "C". Based on the model training, it was tested with test data. If the criteria were not satisfied, the model was trained with the next local and global best positions. The steps of this process are represented in the following steps:

Algorithm 1. PSO with SVM

Input: CIC-DDoS2019 Dataset
Output: Predicted cyberattacks
Step 1: Set the value for ω, cl, c2
Step 2: Initialize the "X" and "V" value
Step 3: Initialize "γ" and "C"
Step 4: Train the model using particle position
Step 5: Analyze the performance of the model with the test dataset
Step 6: If the model satisfies the criteria, the process is stopped or else update particles with new position values and train the model.

In this experiment, the model is considered $\omega = 0.5$, $cl = 1$ and $c2 = 2$. The cognitive velocity is updated with by adding cl and random values similarly social velocity values are updated with by adding ¢2 and random values.

4 Results and Discussion

4.1 Dataset Description

The Canadian Centre for Cybersecurity (CIC) CIC-DDoS2019 databases provided data on normal and malicious network traffic in a realistic test scenario and included genuine network traffic evaluations. A realistic test setup was used to integrate actual network traffic measurements. CICDDoS2019 has 30,480,823 entries, 30,423,960 of which are DDoS attacks and 56,863 benign. The 17.2 GB dataset may be too huge for normal devices to handle.

Table 1 illustrates the performance analysis of proposed PSO with SVM and existing algorithms such as decision tree, random forest, naïve bayes, and SVM.

Table 1. Performance Analysis

Algorithm	Precision	Recall	F1-Score	Accuracy
Random Forest	84	85	84	85
Decision Tree	84	86	85	86
Naïve Bayes	88	89	89	90
SVM	94	95	95	95
Proposed PSO with SVM	95	97	96	97

Figure 1 shows cyber-attack classification performance. The careful examination and selection of algorithms for a specific assignment is crucial. In this case, the PSO with SVM Classifier has the greatest accuracy rate of 97% and is the most successful. SVM Classifier follows closely.

Table 2 shows the DDoS dataset classification report F1-scores for multiple machine learning attacks. The suggested PSO plus SVM method and the SVM algorithm are tested

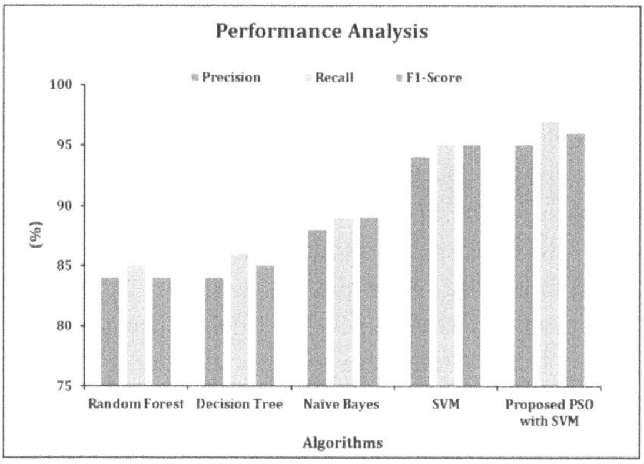

Fig. 1. Performance Analysis

for cyber-attack detection in this research. Certain models showed that the integral of the curve representing benign traffic was higher than the WebDDoS Cyberthreat, which was 0.00. A model should be chosen based on the application's requirements and the risks it faces.

Table 2. Classification - F1-Scores of DDoS Attacks on Proposed PSO with SVM

DDoS Attacks	Random Forest	Decision Tree	Naïve Bayes	SVM	Proposed PSO with SVM
BENIGN	0.98	0.98	0.98	0.98	1
DNS	0.56	0.95	0.94	0.96	0.97
LDAP	0.82	0	0.78	0.99	0.99
MYSQL	0.96	0.98	0.98	0.94	0.98
NetBIOS	0.91	0.76	0.4	0.56	0.95
SSDP	0.05	0	0.65	0.85	0.97
UDP	0.61	0.86	0.91	0	0.99
UDPLag	0.52	0.87	0.92	0.97	0.99
PortMap	0.98	0.84	0.99	0.99	0.99
SYN	0.9	0.97	0.98	0.99	0.98
WebDDoS	0	0	0.07	0.31	0.04

This work has major implications for intrusion detection systems and next-generation firewalls, which protect businesses from cyberattacks. Machine learning approaches can identify DDoS attacks, indicating they might improve intrusion detection systems and next-generation firewalls.

5 Conclusion

This research evaluated DDoS prediction machine learning on the CIC-DDoS2019 dataset. This suggests cybersecurity researchers and practitioners may benefit from the 2019 DDos evaluation dataset. The data showed DDoS attack frequency, duration, and characteristics. Multiple systems and networks could analyze complex DDoS attacks with this version. The 2019 DDos assessment dataset matters for cybersecurity. This sector's primary contribution, the DDoS dataset, will benefit future research. Machine learning may enhance system performance. Next-generation network security research should use AI and ML. These strategies should be utilized to build multiple defenses against various threats. Expanding this endeavor will create a DDoS cyberattack detection and security defense system. The results may help firms strengthen cyber security to prevent DDoS attacks.

References

1. Azzaoui, H., Boukhamla, A.Z.E., Arroyo, D., Bensayah, A.: Developing new deep-learning model to enhance network intrusion classification. Evol. Syst. **13**(1), 17–25 (2022)
2. Birnbaum, D., Borycki, E., Karras, B.T., Denham, E., Lacroix, P.: Addressing publichealth informatics patient privacy concerns. Clin. Govern. Int. J. **20**(2), 91–100 (2015)
3. Yi, H.H., Aye, Z.M.: Performance analysis of traffic classification with machine learning. Int. J. Comput. Inf. Eng. **15**(1), 42–47 (2014)
4. Kurniabudi, Stiawan, D., Darmawijoyo, Bin Idris, M.Y., Bamhdi, A.M., Budiarto, R.: CICIDS-2017 dataset feature analysis with information gain for anomaly detection. IEEE Access **8**, 132911–132921 (2020)
5. Aamir, M., Rizvi, S.S.H., Hashmani, M.A., Zubair, M., Ahmad, J.A.: Machine learning classification of port scanning and DDoS attacks: a comparative analysis. Mehran Univ. Res. J. Eng. Technol. **40**(1), 215–229 (2021)
6. Ambrose, P., Basu, C.: Interpreting the impact of perceived privacy and security concerns in patients' use of online health information systems. J. Inf. Priv. Secur. **8**(1), 38–50 (2015)
7. Cao, Q.W.: Description of SA weak password's harm and solution in the SQL server system. J. Xingtai Polytech. Coll. **29**(1) (2012)
8. Pathinarupothi, R.K.: Clinically aware data summarization at the edge for internet of medical things. In: Proceedings of the IEEE International Conference on Pervasive Computing and Communications Workshops (PerCom Workshops), pp. 437–438 (2019)
9. Raghavan, M.A., Demircioglu, A.: Taeihagh: Public health innovation through cloud adoption: a comparative analysis of drivers and barriers in Japan, South Korea, and Singapore. Int. J. Environ. Res. Public Health **18**(1), 334 (2021). https://doi.org/10.3390/ijerph18010334
10. Salsabila, S., Fikri, D.M., Andika, S.M., Harahap, A.N.: Potential and threat analysis towards cybersecurity in South East Asia. J. ASEAN Dyn. Beyond **1**, 1–13 (2020)
11. Sardi, Rizzi, A., Sorano, E., Guerrieri, A.: Cyber risk in health facilities: a systematic literature review. Sustainability **12**(17), 7002 (2020)
12. Mattei, T.A.: Privacy, con_dentiality, and security of health care information: Lessons from the recent wannacry cyberattack. World Neuro-Surg. **104**, 972–974 (2017)
13. Burke, W., Oseni, T., Jolfaei, A., Gondal, I.: Cybersecurity indexes for eHealth. In: Proceedings of the Australasian Computer Science Week Multiconference, pp. 1–8 (2019)
14. He, Y., Aliyu, A., Evans, M., Luo, C.: Health care cybersecurity challenges and solutions under the climate of COVID-19: scoping review. J. Med. Internet Res. **23**(4) (2021)

15. Hussain, Y.S.: Network intrusion detection for distributed denial-of-service (DDoS) attacks using machine learning classification techniques (2020). dspace.library.uvic.ca
16. Pelletier, Z., Abualkibash, M.: Evaluating the CIC IDS-2017 dataset using machine learning methods and creating multiple predictive models in the statistical computing language R. Science **5**(2), 187–191 (2020)
17. Prakash, K.B.: Quantum meta-heuristics and applications. In: Cognitive Engineering for Next Generation Computing: A Practical Analytical Approach, pp. 265–297 (2021). https://doi.org/10.1002/9781119711308.ch10

AutoProc - AI Based Automated Exam Proctoring System with Test Score Analysis

Sangeeta Parshionikar(✉), Richa Tripathi, Shamita Shetty, Prisha Sharma, and Trupti Shah

Fr. Conceicao Rodrigues College of Engineering, Mumbai, India
sangeeta05@gmail.com

Abstract. Exam Proctoring or invigilation is no longer restricted to a particular time, place or infrastructure. The COVID pandemic caused a fast change in how assessments or evaluations are performed. Proctoring systems currently in use are bulky, cumbersome, and expensive and cannot be used freely in the education sector. In this scenario, an artificial intelligence-assisted proctoring system is needed to monitor online tests that would be most convenient to academic institutions and students. In our work, we developed an AI that enables students to be proctored remotely using the webcam and microphone of their laptops. The proposed framework is separated into three main functional groups - Audio, Visual, and Safe Browser. The remarkable feature of our proctoring system is Test score analysis, which summarises students' malpractices during the entire exam duration.

This particular metric gives a quick output of every student's activity, and the professor can gauge activity through this metric rather than checking the student log for the entire class. It reports multiple face detection with a picture of it, switching of tab to another application for a specific duration or noise detection with its recording. Our efficient approach satisfies the need for an online exam as a trustworthy, helpful tool and supports the expanding online and distance learning sectors/ industries.

Keywords: Auto proctoring · Artificial Intelligence · multiple face detection · object detection

1 Introduction

Over the past few years, significant advancements in the field of education have been made. Educational institutes are looking to technology to give students access to additional resources without impeding their education or jeopardizing their future. Educational institutions require a solution that would be convenient to faculty and students without compromising security, experience, and feasibility.

The demand of the hour is for a web-based intelligent online proctoring service that employs AI-based algorithms to spot dishonest tactics and ensure that examinations are administered correctly. It is possible to guarantee that students won't use unfair or dishonest techniques during the exam by using AI-based remote proctoring and invigilation technologies [1, 2].

V. E. Balas et al. (Eds.): IEQIP 2023, LNNS 1029, pp. 184–190, 2024.
https://doi.org/10.1007/978-3-031-61929-8_27

An AI-based proctor's primary work is utilizing basic computer functionalities like a camera, microphone, and computerized audio and visual gestures to identify plagiarism. Considering the existing proctoring systems, which are bulky and expensive commercial products that cannot be used freely in education, we aim to develop an economical and lightweight proctoring system. By combining manual and AI-based technologies, our AI-based AutoProc prevents cheating as much as possible and maintains the integrity of the exam by developing the necessary implementations. The essential benefits are that there are no physical requirements to be present at an examination center and that it encourages learners to take an exam from any place as long as they meet certain technological conditions.

2 Literature Survey

Nigam, A., Pasricha, R., Singh, T. et al. addressed four primary research issues on the AI-based Proctoring System's existing architecture. The primary concerns included security and privacy concerns, behavioural concerns, a conviction in the current AI-based technology, cost, etc. The discussion also compared a wide range of non-AI-based and AI-based proctoring systems. [1] Y. Atoum et al. proposed the following contributions: They put forward a completely automated method for proctoring online exams that includes optical, i.e., video and audio-based sensors to preserve the uprightness and sincerity of the exam being carried out [2, 3].

H S. G. Asep and Y. Bandung discussed an online exam proctoring system mainly focused on face verification. They proposed a robust method for face verification even in case of variations in poses and lighting without adding much image processing [5]. The authors used the CNN method for face detection and verification. If a single face is detected, then the proposed model verifies it with the user's face. Researchers [6, 7] divided their research papers into four phases. The chosen systems underwent a careful review procedure. The main factors taken into account were the architecture, authorization, final-user support, user identification through verification, regularity of modifications, pricing schemes, security concerns, and more.

The authors highlighted that the examination proctoring procedure in both conventional and online educational programs is the focus of this study's analysis and modelling [8]. The focus is on actions colleges must do to ensure candidate identification, administer the exam, reduce or eliminate cheating, and keep track of exam duration [9]. Authors [10] suggested that further research must be done to integrate cooperative learning strategies with remote proctoring and assessment systems for research plans [11].

3 Methodology and Methods

In this paper, we have developed the proposed system to detect cheating or any malpractices and analyze the students' behaviour during the exam. The various features of the proposed system are identity verification, camera-based capabilities involving mouth movement detection, eye tracking, multiple person detection within camera angle and audio detection. Another salient feature is the Trust score Analysis, which measures the tendency and the frequency of the students to cheat during the test.

3.1 Features of Proposed System

The salient features of the proposed system are as follows:

i. **Verification of Candidate Identity:** The candidate's identity is confirmed before starting the online exam under this proctoring. He/she will be required to login via student Gmail id and password.
ii. **Audio-Based Capabilities:** The system will record audio picked up by the microphone and will be stored in the student logs, which the professor can view after the test ends.
iii. **Mouth Movement Detection:** This is analogous to the detection of eyes. It aims to determine if a person's mouth is open or closed during the exam. The mouth detector detects an infringement when the test taker opens his mouth by measuring the distance between the candidate's lips.
iv. **Eye Tracking:** Eye Movement Detection using AI can help identify if a candidate is gazing straight at the screen, an object, a book, or a cell phone. We track test-takers' eyeballs and see if he's gazing to the left, right, up or down which they might do to signal someone or look around for answers.
v. **Multiple faces and Object Detection:** This part of the system can count the number of persons visible in the webcam to ensure no more than one person is taking the test. Mobile phones, if spotted, are also detected, which helps curb malpractice. Instance segmentation is done to serve this purpose.
vi. **Trust Score Analysis:** Each student will have a trust score calculated. The trust score is the percentage calculation of the summation of students' malpractices in the entire exam duration to the total examination time. The lower the trust score, the greater the chance of students cheating on the test.

The Trust score process (see Fig. 1) is as follows –

a) The default trust score at the start of the test is 100%. As and when the duration of the test decreases, the trust score goes down if any violations are detected
b) Flags are raised for every violation that takes place. The algorithm checks the flag value and calculates the summation of the violations which is deducted from 100 to get a score.
c) Only the professor will be able to see the trust score for an individual student.

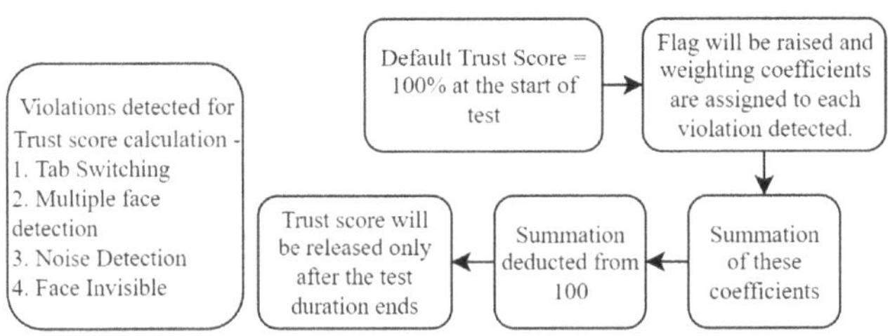

Fig. 1. Trust Score Violation Calculation

3.2 Methodology of the Proposed System

The proposed system is a web-application. It has two main entities: a) Access as a professor - who can create and conduct tests as well as view the results of each student. b) Access as a Student - who can login to the system to view and take the tests conducted by professors.

Access as Professor (see Fig. 2): Initially, to log in as a professor, he/she adds the credentials like name, email, and password. Professors can create examinations, view exam history, and share details of exams with students. The professor can view Live Monitoring of exams and proctoring logs of the individual student.

1. Once the professor successfully logs in on the platform, the next step would be creating a test/exam link using Google forms, add the questions and post the link.
2. The test can be enabled according to the scheduled time, by the professor.
3. For stringent execution of the entire examination, the professor can act as a human proctor in a meeting for monitoring the students.
4. Once the test ends, the professor can view each and every student's activity log on the portal. The professor will have the provision of setting a threshold of 75% trust score and segregate students below 75% trust score.

Access as a Student (see Fig. 3):

1. Once the test link is posted and enabled by the professor, students need to log in through a Gmail account using their student's credentials.
2. After successful login, a pop-up asks for mic and camera permissions. The students are asked to share their entire screen in order to monitor their activity.
3. During the test, students' gestures and activities are detected, and a snapshot is taken after every 7 s of the interval.
4. The system monitors students' activity throughout the test on the parameters of the features mentioned above.

4 Results and Discussion

4.1 Face Detection

For Face Detection, we used a Haar Cascade classifier. It is an object detection algorithm used to identify faces in an image or a real-time video. The Face Detection part of the project was successfully (see Fig. 4).

4.2 User Verification

The welcome page has register and log in. When the user visits the page for the first time, they must register and fill out the credentials, then proceed to the login page. From now on, users must log in when they want to take the exam. Figure 5 shows a page where the teacher can create a test link, and Report of Submissions (see Fig. 6), the test score results (see Fig. 7).

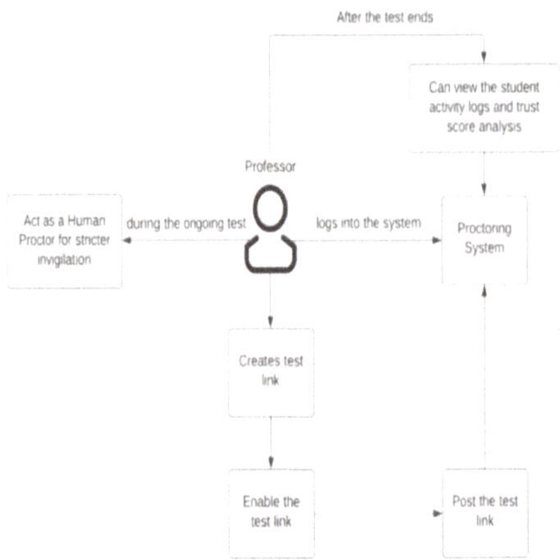

Fig. 2. Access as a professor

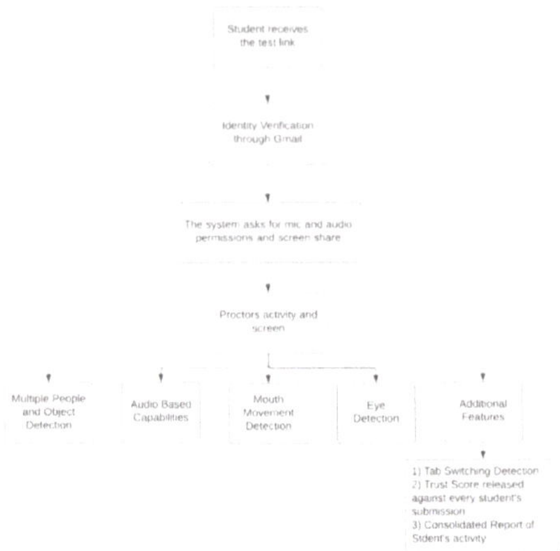

Fig. 3. Access as a student

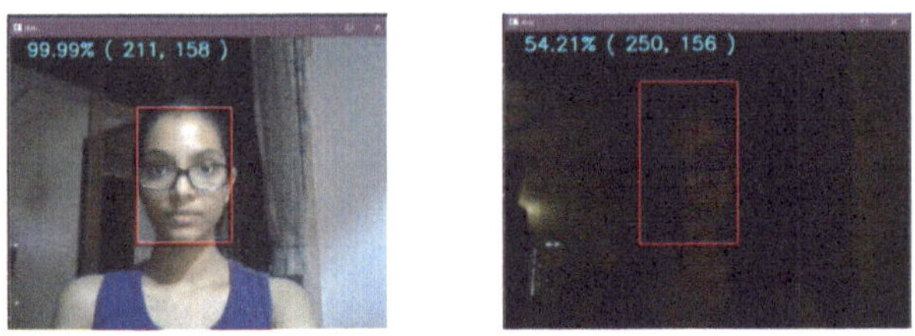

Fig. 4. Face Detection in normal light and in dim lights

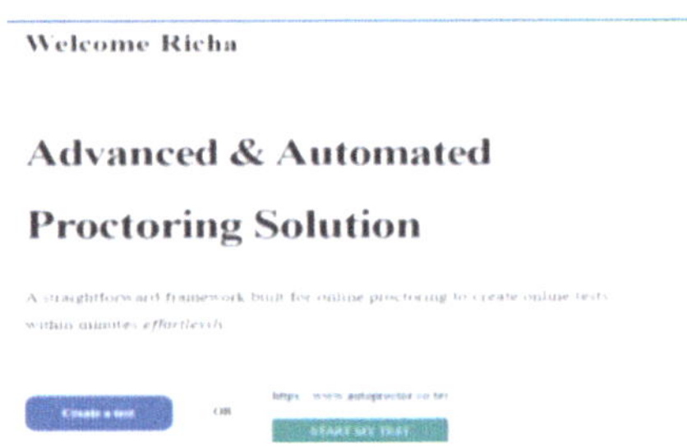

Fig. 5. Create Test Link

Fig. 6. Report of Submitted Papers

Fig. 7. Trust Score Results

5 Conclusion

The importance of digital education is understood and accepted all over the word. To prevent cheating to a fair extent and maintain the integrity of the exam, specific vision and audio-based features are implemented in our AutoProc system. Our proctoring model detects face in all different angles along with eye and mouth movements. Trust score analysis reported malpractices such as multiple face detection with picture of it, switching of tab to other application for specific duration or noise detection with its recording. With the dependence of this framework, we can now conclude that the proctoring system helped in the authorization of the institutions to maintain the security and integrity of its examinations' data like records of attendance, attempts, exam results, etc. Artificial intelligent models screen and follow the whole framework, which are precise and unbiased.

References

1. Nigam, A., Pasricha, R., Singh, T., et al.: A systematic review on AI-based proctoring systems: past, present and future. Educ. Inf. Technol. **26**(5), 6421–6445 (2021)
2. Atoum, Y., Chen, L., Liu, A.X., Hsu, S.D.H., Liu, X.: Automated online exam proctoring. IEEE Trans. Multimed. **19**(7), 1609–1624 (2017)
3. Jia, J., He, Y.: The design, implementation and pilot application of an intelligent online proctoring system for online exams. Interact. Technol. Smart Educ. **19**(1), 112–120 (2022)
4. Chandra, N., et al.: Automating Online Proctoring Through Artificial Intelligence. Int. Res. J. Eng. Technol. (IRJET) **08**(01), 1894–1896 (2021)
5. Asep, H.S.G., Bandung, Y.: A design of continuous user verification for online exam proctoring on M-learning. In: International Conference on Electrical Engineering and Informatics (ICEEI), Bandung, Indonesia, pp. 284–289 (2019)
6. Yadav, S., Singh, A.: An image matching and object recognition system using webcam robot. In: Fourth International Conference on Parallel. Distributed and Grid Computing (PDGC) (2016)
7. Prathish, S., Athi Narayanan, S., Bijlani, K.: An intelligent system for online exam monitoring. In: International Conference on Information Science (ICIS) (2016)
8. Kaddoura, S., Gumaei, A.: Towards effective and efficient online exam systems using deep learning-based cheating detection approach. Intell. Syst. Appl. **16**, 200153 (2022). https://doi.org/10.1016/j.iswa.2022.200153. ISSN 2667-3053
9. Turani, A.A., Alkhateeb, J.H., Alsewari, A.A.: Students online exam proctoring: a case study using 360 degree security cameras. In: 2020 Emerging Technology in Computing, Communication and Electronics (ETCCE), Bangladesh, pp. 1–5 (2020). https://doi.org/10.1109/ETCCE51779.2020.9350872

10. Sandhya, G., Rao, V.Y., Rawat, N., Chowdary, S., Satish, D.H.: Online proctoring system "ProctR." Int. J. Res. Publ. Rev. **4**(7), 470–475 (2023)
11. Khan, M., Chakraborty, S., Astya, R., Khepra, S.: Face detection and recognition using OpenCV (2019). https://doi.org/10.1109/ICCCIS48478.2019.8974493

Network Summarization Through the Lens of Latent Semantic Analysis and Fuzzy Techniques

M. Vanathi(✉) and L. Lakshmanan

Department of Computer Science and Engineering, Sathyabama Institute of Science and Technology, Chennai, India
{vanathi.cse,lakshmanan.cse}@sathyabama.ac.in

Abstract. Extracting text summarization involves the retention of only the most important sentences in a document. Finding the appropriate ranking order of the phrases in the document according to their relevance is a vital step in extractive text summarization. Singular value decomposition or SVD algorithm based on latent semantic analysis focuses on recognizing the sections in the document which are related in terms of their semantic nature. Fuzzy algorithms involve reasoning of the priority order of the sentences using fuzzy logic unlike the use of discrete values. While significant work has been done for extractive text summarization in English and other foreign languages, there is ample scope for improving the performance of systems when dealing with Marathi text. An analysis of the characteristics of both these techniques is conducted to compare their benefits and shortcomings. The performance of both the algorithms is evaluated on a document dataset using standard performance metrics including the ROUGE metric. Both of these approaches are objectively compared in order to determine their applicability, particularly when dealing with non-English text in general or Marathi in particular.

Keywords: Extractive text summarization · fuzzy logic · Marathi text · Singular value decomposition

1 Introduction

Natural language processing involves processing and modeling of natural language data to ameliorate understanding of computers to make sure that the semantic and syntactic structure of the data is retained. Extractive text summarization is a summarization type where there is no addition of new content or modification of existing content, but rather only the most important phrases and sentences from the document are retained as the summary of document [1, 2]. Based on a decided threshold, the top N ranked sentences would then be predicted as the summary of the document. Previous approaches in this domain have considered the use of statistical features such as word count, and term frequencies for ranking the sentences [3]. The English language has traditionally been considered for demonstrating and conceptualizing extractive text summarization systems [4]. The

V. E. Balas et al. (Eds.): IEQIP 2023, LNNS 1029, pp. 191–197, 2024.
https://doi.org/10.1007/978-3-031-61929-8_28

research community's contributions to non-English languages, particularly numerous Indian languages, have increased in recent years [2]. Marathi is a language spoken in the state of Maharashtra and nearby regions in India and is derived from the Devanagari script [2]. The work on extractive text summarization in the Marathi language has been relatively paltry [5–7]. In this paper, the purpose of extractive text summarization on Marathi text documents, two distinct methods are suggested. The first approach focuses upon using latent semantic analysis (LSA) that deploys a semantic identification and correlation of sentences between a document [8]. The second approach makes use of fuzzy logic to solve this task. The fuzzy algorithm focuses on certain text characteristics and rules [9] using which sentence scores are assigned. The contributions in this paper has made: Text summarization technique on LSA, implementation of fuzzy logic and analysis and comparison of the aforementioned two approaches to better guide further work in this domain. The remaining information is organized as follows: former work is bandied in Sect. 2, the two proposed approaches are described in detail in Sect. 3, the results and analysis are carried out in Sect. 4 and the conclusion in Sect. 5.

2 Related Works

Extractive text summarization has been the preliminary text summarization demonstrating the priority ranking capability of the data modeling algorithm. The initial method for extractive text summarization made use of term frequency and inverse document frequency for feature selection [3]. Kumar et al. [11] made use of a knowledge induced graph for performing single document summarization. In the last decade, machine learning has also been used to tackle this task across multiple domains as well as multiple languages [12, 13]. Query- based text summarization has also been tried where the ranking is based on the overlapping between the query phase and the document terms [14]. Recently, there have been attempts to perform extractive text summarization in Marathi. Bhosale et al. [7] used a naive frequency count approach for this task. Rathod made use of the page-rank and text-rank algorithms, however, their evaluation of these approaches was very restricted [15]. Sarwadnya and Sonawane went a step further in terms of the use of preprocessing methods and reliance on the text-rank algorithm [5]. Chaudhari et al. [6] presented the use of deep learning to create the summarizer. Fuzzy logic and singular value decomposition [1, 16], there hasn't been much community involvement in applying these to Marathi documents. In this paper, an attempt is made to try to incorporate these approaches in a novel way to boost performance when handling documents written in Marathi.

3 Proposed Methodologies

A. Preprocessing: Preprocessing steps of the input documents is followed by the removal of stop words, stemming, tokenization. Marathi language is characterized by the addition of suffixes to verbs to indicate the gender or the tense in which the sentence is being spoken.
B. Summarizer algorithms: Two different algorithms are presented in this paper for extractive text summarization. These are as follows: Singular Value Decomposition (SVD), Fuzzy Logic

Singular Value Decomposition (SVD): Computing the latent semantic structure of the document to obtain context similarity between the sentences and thereby mapping the vector space. Algorithm 1 indicates the procedure to derive the SVD values for subsequent ranking of the sentences in the document.

Algorithm 1 Algorithm for computing SVD
Input: Normalized input representation matrix D
Output: Values of U, V, and Σ
1: $Prod\_D = DDT$
2: $X1 = Eigen\_values(Prod\_D)$
3: $Inv\_D = DTD$
4: $X2 = Eigen\_values(Inv\_D)$
5: $Val = \sqrt{x1} \cap x2$
6: *Assign Values to U, V, and Σ*
7: Return U, V, Σ

As a modification to the existing SVD approach, three other factors are also considered. Each of these factors is considered and evaluated as follows:

- Sentence length: If the sentence length is less than the minimum permissible length, or greater than the maximum permissible length, then set it to zero.
- Sentence value: The previously discussed normalized input representation
- Sentence position: If the sentence appears first or last in the document, it is deemed important and is assigned a value of 1. Otherwise, derive the value as follows:

$$sentence\,pos = cos\frac{Pos - (TRSH * len(sentence)) * 360}{(1 - 2 * TRSH)len(sentence)}$$

  where TRSH is a hyperparameter decided by the user. The value is set to 0.01 in the presented setup.
- Sentence weight similarity: Calculated using the number of overlapping words present between two sentences.

The final ranking for the sentence is derived by considering the sum of the absolute values of each of these factors. Based on the summary factor given by the user, the ranked sentences are sorted in descending order and the filtered sentences are output as the summary of the document.

Fuzzy Logic: The proposed fuzzy logic is calculated using a feature matrix. The feature matrix is derived based upon certain statistical features present in the document. Each of these features is as follows:

- Position factor of the sentence: The position factor of the sentence is calculated by normalizing its order in the text with respect to the overall number of sentences.

$$pos\,factor = \frac{Total\_sentence - current\_pos}{Total\_sentence}$$

- Bigram token length: Bigram is the tokenization of words done, but by considering two words at a time. Trigram token length: Trigrams are similar to bigram, but they consider three words together at a time.

- TF-ISF vector: It considers the term frequency as well as sentence frequency and is calculated as follows:

$$tf\_isf = \frac{term\_freq}{sent\_freq * vocab\_pos}$$

- Cosine similarity: Calculate the cosine similarity of the sentence with respect to the centroid of the document.

$$Cos\_similarity(S, Z) = \frac{S.Z}{\|S\|^2 \cdot \|Z\|^2}$$

where Z is the centroid of the document and S is the sentence in consideration.
- Thematic number: It takes into consideration the factor of the number of keywords present in a sentence with respect to the total keywords present in the document [9].

$$Thematic\_number(s) = \frac{keywords\ in\ S}{total\ keywords}$$

The fuzzy logic requires a triangular membership function used to control the shape of the function [19]. Based on the previously mentioned nine factors, a consequent factor is determined that is termed as bad, average, and good for vector values of [0, 0, 50], [0, 50, 100], and [50, 100, 100] respectively. For an instance of data, the values of the aforementioned nine factors are calculated per sentence and passed as input for the fuzzy logic to compute.

4 Result and Analysis

The performance of the two proposed approaches is evaluated on a custom created dataset consisting of Marathi news articles ranging on a diverse set of issues including politics, economics, and social affairs. As both the methods are instance-based and do not involve any trainable parameters, the entire dataset consisting of all the 100 documents is used for evaluation purposes. Extractive text summarization focuses on retention of the most important sections of the document. As a result, evaluation of such summarizers focuses on the amount of overlap between the human summary and the machine generated summary. To define this measure of overlap in a standard format, the ROUGE metric is used [20]. Based on the performance metrics mentioned, the results are produced and tabulated in Table 1.

Table 1. Results obtained on both the approaches for single document summarization

Metric	SVD	Fuzzy Logic
ROUGE1: Precision	0.632	**0.641**
ROUGE1: Recall	0.623	**0.625**

(continued)

Table 1. (*continued*)

Metric	SVD	Fuzzy Logic
ROUGE1: F1	0.612	**0.623**
ROUGE2: Precision	0.531	**0.561**
ROUGE2: Recall	0.519	**0.546**
ROUGE2: F1	0.512	**0.546**
ROUGEL: Precision	**0.665**	0.659
ROUGEL: Recall	0.614	**0.636**
ROUGEL: F1	0.626	**0.64**

It can be seen that the Fuzzy logic turns out to be a better approach as compared to the SVD method with better results on almost all of the performance metrics. Next, multi-document summarization is considered. In this case, the overlaps of tokens in the human summary and machine-generated summary are considered across multiple documents. Evaluation is done for precision, recall, and the F1-score. The results obtained are shown in Table 2.

Table 2. Results obtained on both the approaches for multi-document summarization

Metric	SVD	Fuzzy Logic
Precision	**0.705**	0.625
Recall	**0.693**	0.655
F1	**0.682**	0.63

It can be seen that while SVD was lagging in case of single document summarization, it outperforms Fuzzy logic by a comfortable margin when it comes to multidocument summarization. Further, as a part of ablation studies, the F1 scores are compared with a recently used method of position-based text rank [20].

5 Conclusion

In this paper, two novel approaches have been proposed for the task of extracting the text summarization on Marathi documents. The initial strategy focused on singular value decomposition as a dimensionality reduction and feature selection technique. The second approach made use of fuzzy logic to derive rules used for priority ranking of sentences based on certain statistical features in the document. Both the proposed approaches have certain advantages and shortcomings. The evaluation of the approaches was done on a standard-sized dataset and a fuzzy logic-based approach was found to be better when working on single document classification. On the other hand, SVD was seen to be the better method for multi document summarization. There is a certain accuracy complexity tradeoff amongst the two approaches. This has been demonstrated by the evaluation of multiple performance metrics. Future scope in this domain includes consideration of semantic analysis, word embeddings, and extension of the task to include abstractive text summarization. The proposed approaches have shown promising signs for text summarization tasks in Marathi language and could be extended further to other natural language understanding tasks.

References

1. Gupta, V., Lehal, G.S.: A survey of text summarization extractive techniques. J. Emerg. Technol. Web Intell. **2**(3), 258–268 (2010)
2. Giri, V.V., Math, M., Kulkarni, U.: A survey of automatic text summarization system for different regional language in India. Bonfring Int. J. Softw. Eng. Soft Comput. **6**, 52–57 (2016). Special Issue Special Issue on Advances in Computer Science and Engineering and Workshop on Big Data Analytics Editors: Dr. SB Kulkarni, Dr. UP Kulkarni, Dr. SM Joshi and JV Vadavi
3. Kaur, J., Gupta, V.: Effective approaches for extraction of keywords. Int. J. Comput. Sci. Issues (IJCSI) **7**(6), 144 (2010)
4. Kaikhah, K.: Text summarization using neural networks (2004)
5. Sarwadnya, V.V., Sonawane, S.S.: Marathi extractive text summarizer using graph-based model. In: 2018 Fourth International Conference on Computing Communication Control and Automation (ICCUBEA), pp. 1–6. IEEE (2018)
6. Chaudhari, A., Dole, A., Kadam, D.: Marathi text summarization using neural networks. Int. J. Adv. Res. Dev. **4**(11), 1–3 (2019)
7. Bhosale, M.S., Joshi, D., Bhise, M.V., Deshmukh, R.A.: Automatic text summarization based on frequency count for marathi e-newspaper (2018)
8. Badry, R.M., Eldin, A.S., Elzanfally, D.S.: Text summarization within the latent semantic analysis framework: comparative study. Int. J. Comput. Appl. **81**(11), 40–45 (2013)
9. Yadav, J., Meena, Y.K.: Use of fuzzy logic and wordnet for improving performance of extractive automatic text summarization. In: 2016 International Conference on Advances in Computing, Communications and Informatics (ICACCI), pp. 2071–2077. IEEE (2016)
10. Deshpande, A.R., Lobo, L.: Text summarization using clustering technique. Int. J. Eng. Trends Technol. **4**(8), 3348–3351 (2013)
11. Kumar, N., Srinathan, K., Varma, V.: A knowledge induced graph-theoretical model for extract and abstract single document summarization. In: Gelbukh, A. (ed.) CICLing 2013. LNCS, vol. 7817, pp. 408–423. Springer, Heidelberg (2013). https://doi.org/10.1007/978-3-642-37256-8_34
12. Sarkar, K., Nasipuri, M., Ghose, S.: Using machine learning for medical document summarization. Int. J. Database Theory Appl. **4**(1), 31–48 (2011)
13. Bazrfkan, M., Radmanesh, M.: Using machine learning methods to summarize persian texts. Indian J. Sci. Res. **7**(1), 1325–1333 (2014)
14. Imam, I., Nounou, N., Hamouda, A., Allah, H., Khalek, A.: Query based Arabic text summarization 1 (2013)
15. Rathod, Y.V.: Extractive text summarization of Marathi news articles (2018)
16. Patil, P.D., Kulkarni, N.: Text summarization using fuzzy logic. Int. J. Innov. Res. Adv. Eng. (IJIRAE) **1**(3), 42–45 (2014)

17. Steinberger, J., Ježek, K.: Text summarization and singular value decomposition. In: Yakhno, T. (ed.) ADVIS 2004. LNCS, vol. 3261, pp. 245–254. Springer, Heidelberg (2004). https://doi.org/10.1007/978-3-540-30198-1_25
18. Suanmali, L., Binwahlan, M.S., Salim, N.: Sentence features fusion for text summarization using fuzzy logic. In: 2009 Ninth International Conference on Hybrid Intelligent Systems, vol. 1, pp. 142–146. IEEE (2009)
19. Kyoomarsi, F., Khosravi, H., Eslami, E., Dehkordy, P.K., Tajoddin, A.: Optimizing text summarization based on fuzzy logic. In: Seventh IEEE/ACIS International Conference on Computer and Information Science (ICIS 2008), pp. 347–352. IEEE (2008)
20. Lin, C.-Y., Och, F.: Looking for a few good metrics: rouge and its evaluation. In: Ntcir Workshop (2004)
21. Mihalcea, R., Tarau, P.: TextRank: bringing order into text. In: Proceedings of the 2004 Conference on Empirical Methods in Natural Language Processing, pp. 404–411 (2004)

Acceleration of Variational Quantum Eigen Solver

Manish Modani(✉)

NVIDIA Graphics Pvt Ltd., Pune 411006, Maharashtra, India
mmodani@nvidia.com

Abstract. Hybrid algorithm, Variational Quantum Eigen (VQE) solver plays key role in drug discovery, material science, chemical engineering, portfolio optimization, artificial intelligence etc. The accelerated version of VQE, implemented using NVIDIA cuQuantum SDK and executed on NVIDIA DGX-A100 system. The VQE accelerated algorithm performance compared with its CPU only implementation, for 25, 28, 30 and 34 qubits respectively. The run time increases multifold, with the increase of qubits, when executed on CPU, however for GPU enabled run, the execution time increases marginally. The relative speed up between CPU only and GPU enabled run, for one stable run, increases from 37× (25 qubits) to 951× (34 qubits). In practice, VQE is executed, number of times, until the minimum expectation value of Hamiltonian is obtained, by varying the ansaz parameters. The speed up between, CPU only and GPU enabled run, becomes significant with the increase in the iterations of VQE execution.

Keywords: Variational Quantum Eigen Solver (VQE) · Quantum Computing Simulation · Accelerated Computing · Cirq · A100 GPU · NVIDIA

1 Introduction

Quantum computing, leveraging the quantum mechanical properties of superposition and entanglement, has undergone meteoric development in the recent past. Quantum computers with 100's of qubits, are available and accessible. Recently, IBM unveiled superconducting based 433 qubit osprey system [1]. Pino et al. [2] demonstrated the path for Quantum Computers using trapped-ion quantum charge-coupled device (QCCD). The quality of quantum circuits on superconducting quantum computing systems is also improved by Jurcevic et al. [3]. Though there is significant progress, yet the Quantum technology is in noisy intermediate-scale quantum (NISQ) regime [4]. It limits the Quantum Algorithm applicability and performance. The Quantum Computing Simulation (QCS) on classical computers plays vital role in quantum algorithms development and implementation, validation, simulation of upcoming quantum architecture etc. Top 63 Quantum computing simulators in 2022 are summarized by Dargan [5]. However, most of these simulators are either limited by number of qubits (memory) or performance.

Peruzzo et al. [6] developed the Variational Quantum Eigen (VQE) Solver. Its theoretical framework is given by McClean et al. [7]. Variational principle is used in VQE

V. E. Balas et al. (Eds.): IEQIP 2023, LNNS 1029, pp. 198–205, 2024.
https://doi.org/10.1007/978-3-031-61929-8_29

to compute Hamiltonian's ground state energy, a critical problem in the study of quantum chemistry [8–10], condensed matter physics [11, 12], optimization [13] and in artificial intelligence [14, 15]. It results, VQE wide uses in drug discovery [16, 17], material science [18], chemical engineering [19], portfolio optimization [20] and reinforcement learning [21] etc. Recently, Tilly et al. [22] and Fedorov [23] summarized VQE uses, best practices and advantages/limitations associated with the various form of its implementation with QCS.

Accelerated computing is the use of specialized hardware (eg NVIDIA GPU) to speed up its execution time or throughput or both. NVIDIA cuQuantum SDK, optimized libraries and tools, accelerate the performance of quantum computing workflows based on state vector as well as tensor network methods [24]. In the present work, the commonly used hybrid variational eigne solver (VQE) for 2D Ising model with transverse field is simulated on NVIDIA DGX A100 system, which enabled the simulations for larger number of qubits (upto 36). The performance advantage in comparison to CPU only run are also demonstrated. In the next section, VQE algorithm is explained. It is followed by problem statement, experimental setup, results and discussions sections. Finally, findings are summarized in the conclusions section.

2 Problem Statement

The VQE algorithm runs partly on a classical computer and quantum processing units (QPU).

An instance of VQE has two algorithmic sub-components (i) a trial state/ansatz and (ii) the classical optimizers. Classical Optimizer varies the ansatz (Quantum Circuit) via set of parameters, such that it results in the minimum expectation value (Fig. 1).

In this work, the classical NP-complete problem, 2D Ising model with transverse field is solved. The problem details are given at [25]. However, for the completeness, the problem and its implementation are briefly discussed below.

Consider the energy function.

$$E(s_1, s_2, \ldots .s_n) = \sum\nolimits_{(i,j)} J_{i,j} s_i, s_j + \sum\nolimits_i h_i s_i \quad (1)$$

here each s_i $J_{i,j}$ and h_i valuers are +1 or −1. Index i is associated with a bit on a square lattice. The $\langle i, j \rangle$ notation represents the sum over neighboring bits on the lattice. Here one needs to find s_i values to minimize E for the given $J_{i,j}$ and h_i. In variational quantum algorithm, the classical and quantum problem are mapped by considering n-qubits (quantum) and associate them with bits (classical) to minimize the expectation value.

$$H = \sum\nolimits_{(i,j)} J_{i,j} Z_i, Z_j + \sum\nolimits_i h_i Z_i \quad (2)$$

Then one defines a set of parameterized quantum circuits. In this, the quantum circuits are assigned the values. It results an ansatz state $|\varphi(p_1, p_2, \ldots, p_k)\rangle$. Where p_i produces this state. The steps to obtain value of objective function using variational algorithm are summarized below:

I. Ansatz state preparation.
II. Execution on QPU/Simulator and measurement.
III. Classical optimizers check the convergence. In case of no convergence, Goto (I).

Step (II) repeated many times to have stable solution for the expectation value. Variation on a Quantum Approximate Optimization Algorithm (QAOA) strategy followed to define an ansatz related to the given problem.

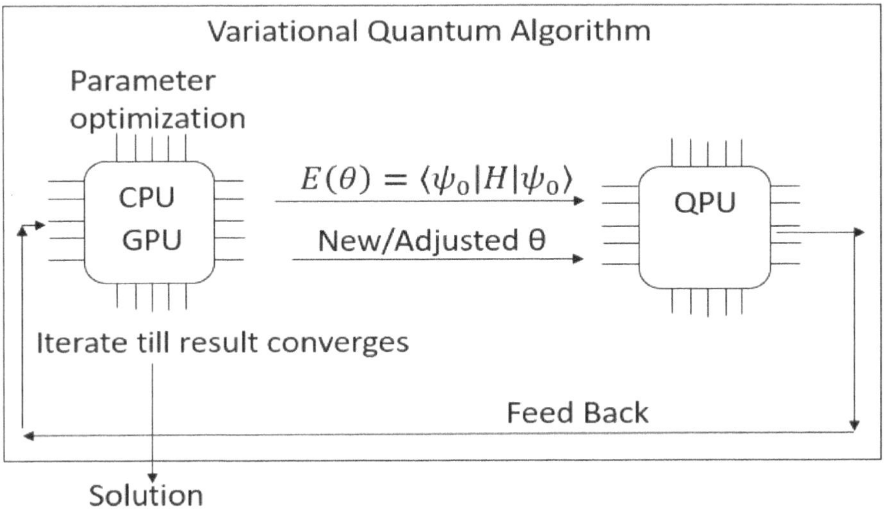

Fig. 1. Hybrid HPC-QC algorithm architecture

3 GPU Acceleration

In Quantum Circuit Simulation, numerical quantum computation is performed on a classical computer. The naive approach is limited in terms of size, as with the increase in problem size (qubits), the memory requirement scales exponentially. NVIDIA cuQuantum SDK [24] provides composable primitives for GPU-accelerated quantum circuit simulations. NVIDIA cuQuantum Appliance is multi-node and multi-GPU enabled solution, provides optimization for state vector and tensor network simulations using cuStateVec and cuTensorNet libraries respectively. As part of cuQuantum SDK, VQE is accelerated and integrated with almost all the available frameworks (Qiskit, Cirq etc.). For illustration, here we consider the implementation of VQE using Cirq framework.

Experiment Setup:
For the performance advantage illustration, the 2D Ising Hamiltonian problem, implemented using Cirq framework is used. 2D Ising Hamiltonian problem is solved using VQE to find the minimum energy. The code for random Ising Hamiltonian on a rectangular grid is developed. It defines a few helper functions to generate VQE ansatz. The simulation runs multiple times on QPU/Simulator to have stable solution. Then the

output goes in the classical optimizer, where it compares the obtained energy with the given threshold. If the obtained energy is higher than the given threshold, the classical optimizer changed the set of input parameters, pass it over to QPU/Simulator, new ansatz is formed, and simulations are performed again to have optimal solution.

In this application, the values Ji,j and *h*i are randomly assigned to +1 or −1. The ansatz starts in the all plus state and it is evolved QAOA-type approach. The "mixing" term added by 0.3 in the "x layer" and "z layer". It is entangled using controlled z (CZ) gate. In case of J = +1, CZ applied whereas in case of J = −1, both the qubits were flipped and then CZ gate applied. These helper function generated the ansatz.

Simulations are performed using cirq.Simulator or qsimcirq.QSimOptions functions for CPU and GPU enabled run respectively. Simulations repeated 100 times to have the stable solution. The energy expectation for the given ansatz is calculated to find the minimum energy for the given Hamiltonion. The objective function's value for the initialized ansatz is observed as 13.28. As a next iteration, the ansatz is parameterized by sweeping over three angles (eg: alpha, beta, and gamma). Values of these angles varied from 0 to 1. Sweeping over 10 angles (from 0 to 1) for each parameter gives us 1000 circuits to run. For 25 qubits, the observed minimum state is −3.04. The experiment repeated for GPU enabled simulation and similar results are observed (eg for first time initialized ansatz, the observed energy was 13.74 and after ansatz parameterization by sweeping all the three angles, the observed energy is −3.22). It shows that the results (minimum energy) observed from the GPU enabled VQE are almost same.

The simulation time, to compute and measure the energy, is considered to demonstrate the performance enhancement. Here, to have stable solution, the simulation are repeated 100 times. The compute intensive part e.g. simulation are offloaded to GPUs. The function, cirq.Simulator and qsimcirq. QSimSimulator with option disable_gpu = True is used to observe the CPU only compute time. The function qsimcirq.qsimsimulator with disable_gpu = False is used to observe the run time for GPU enabled simulation. Qsimcirq.qsimsimulator is the optimized routine to execute simulation with various option (eg number of threads, gpu usage enabled/disabled, number of GPUs etc.). Hence, for CPU only run, the execution time observed by qsim_simulator.run routine is better (less) than that observed by simulator.run. Therefore, for further analysis, CPU and GPU enabled run times are observed by qsim_simulator.run are considered. The time taken by 25, 28, 30 and 34 qubits are shown in Table 1. The run for 36 qubits also performed. The execution time for GPU only run is observed as 8.9 s, however, the CPU run was taking very long, hence it is not included for the analysis.

Number of runs (minimum three) were made by varying parameters eg number of CPU threads, number of GPUs. For comparison purposes the best performance observed from CPU only run and GPU enabled run are considered. NVIDIA DGX-A100 80 GB system (which has 128 cores, AMD EPYC 7742 the system and 8 A100 GPUs with 80 GB memory of each GPU) is used for the experiments. As an illustration, execution on 8 GPUs and its usage are shown below Fig. 2.

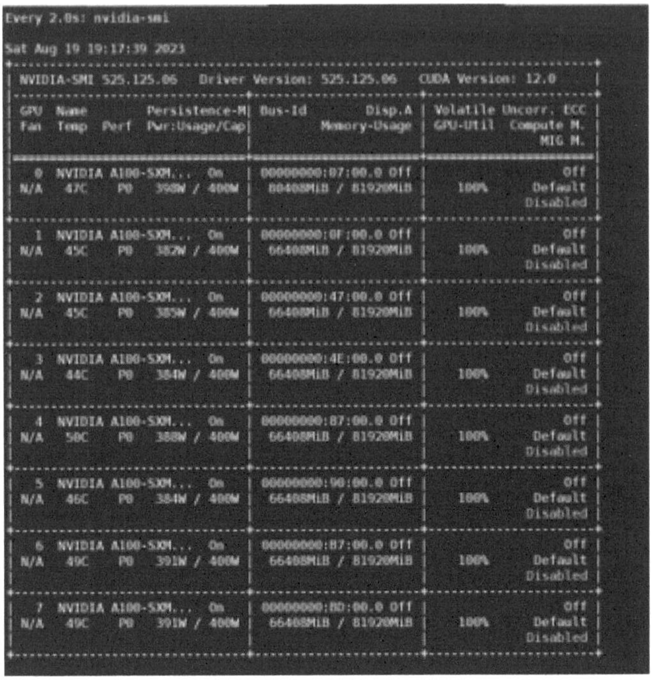

Fig. 2. Demonstration of GPU usage

The execution time taken by CPU run and gpu enabled run are given in Table 1.

Table 1. Execution time in Seconds

Qubits	CPU Time (Seconds)	GPU Time (Seconds)
25	1.9	0.05
28	19.1	0.38
30	71.91	0.50
34	1559.8	1.64

The execution time for CPU only run increases significantly with the increase of number of qubits. It could be due to the fact, the memory requirements increase with the Increase of qubits. It could result in page faults, which impacts overall performance negatively. In contrast, the cuQuantum SDK is optimized to exploit the GPUs capabilities, which results marginally increase in execution time with the increases in the qubits. For 34 qubits, the observed speed up is 951× (Fig. 3).

In the present work the acceleration of VQE performance is demonstrated using NVIDIA cuQuantum SDK and GPUs. These needs to be further analyzed with different

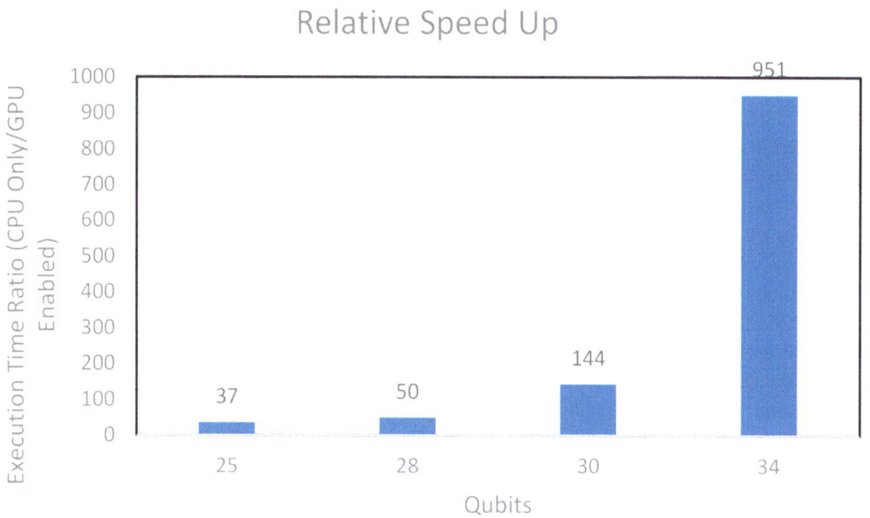

Fig. 3. Performance advantage by GPU enabled run over CPU only run for 25 to 34 qubits.

kind of ansatz parameterization. In addition, the acceleration for real time application also needs to be explored.

4 Conclusions

The variational method in quantum theory is a hybrid method, which runs on both classical and quantum processing units, to find the low energy states of a quantum system. It is commonly used in drug discovery, material science, chemical engineering, portfolio optimization, artificial intelligence etc. In view of the limitations associated with the available QPU, Quantum Simulations on classical systems plays vital role in algorithm development and analysis. In this work, the acceleration of VQE algorithm on GPUs are demonstrated. For demonstration, VQE implementation with Cirq framework is considered. The CPU only version of Cirq takes very long for larger (>30) number of qubits. The accelerated version of VQE is implemented using NVIDIA cuQuantum SDK and executed on DGX-A100 system. The VQE accelerated algorithm for 2D Ising Hamiltonian problem could run for 36 state vector qubits, however for the comparison purposes, the performance between CPU only and GPU enabled VQE is demonstrated for 25, 28, 30 and 34 qubits respectively. The execution time for CPU run increases significantly with the increase in the number of qubits however, for the GPU run, the execution time increases marginally. Accordingly, the relative speed up between CPU and GPU enabled run, increases from 37× (25 qubits) to 951× (34 qubits). It should be noted, this speed up is obtained by comparing the execution time of one simulation. The speedup will increase multi-fold with the ansaz parameterization to find the minimum energy state.

Acknowledgements. Author would like to thank Takuma Yamaguchi, NVIDIA for his help in code modifications.

References

1. https://www.ibm.com/quantum
2. Pino, J.M., et al.: Demonstration of the trapped-ion quantum CCD computer architecture. Nature **592** (7853), 209–213 (2021). https://www.nature.com/articles/s41586-021-03318-4
3. Jurcevic, P., et al.: Demonstration of quantum volume 64 on a superconducting quantum computing system. Quantum Sci. Technol. **6**(2), 025020 (2021). https://doi.org/10.1088/2058-9565/abe519
4. Wendin, G.: Quantum information processing with superconducting circuits: a perspective. Reference modules in Materials Science and Material Engineering (2023). https://arxiv.org/abs/2302.04558
5. https://thequantuminsider.com/2022/06/14/top-63-quantum-computer-simulators-for-2022/
6. Peruzzo, J., et al.: A variational eigenvalue solver on a photonic quantum processor. Nat. Commun. **5**(1) (2014) https://doi.org/10.1038/ncomms5213
7. McClean, J.R., Romero, J., Babbush, R., Aspuru-Guzik, A.: The theory of variational hybrid quantum-classical algorithms. New J. Phys. **18**(2), 023023 (2016). https://doi.org/10.1088/1367-2630/18/2/023023
8. Deglmann, P., Schäfer, A., Lennartz, C.: Application of quantum calculations in the chemical industry-an overview. Int. J. Quantum Chem. **115**(3), 107–136 (2014). https://doi.org/10.1002/qua.24811
9. Williams-Noonan, B.J., Yuriev, E., Chalmers, D.K.: Free energy methods in drug design: prospects of "alchemical perturbation" in medicinal chemistry. J. Med. Chem. **61**(3), 638–649 (2017). https://doi.org/10.1021/acs.jmedchem.7b00681
10. Heifetz, A. (ed.): Quantum Mechanics in Drug Discovery. Springer, New York (2020). https://doi.org/10.1007/978-1-0716-0282-9. ISBN 9781071602829
11. Continentino, M.A.: Key Methods and Concepts in Condensed Matter Physics. IOP Publishing, UK (2021). ISBN 9780750333955
12. der Ven, A.V., Deng, Z., Banerjee, S., Ong, S.P.: Rechargeable Alkali-ion battery materials: theory and computation. Chem. Rev. **120**(14), 6977–7019 (2020). https://pubs.acs.org/doi/epdf/10.1021/acs.chemrev.9b00601
13. https://www.classiq.io/insights/quantum-algorithms-vqe
14. Choi, J., Oh, S., Kim, J.: The useful quantum computing techniques for artificial intelligence engineers. In: 2020 International Conference on Information Networking (ICOIN), Barcelona, Spain, pp. 1–3 (2020). https://ieeexplore.ieee.org/document/9016555
15. Nakayama, A., Mitarai, K., Placidi, L., Sugimoto, T., Fujii, K.: VQE-generated quantum circuit dataset for machine learning (2023). https://arxiv.org/abs/2302.09751
16. Cao, Y., Romero, J., Aspuru-Guzik, A.: Potential of quantum computing for drug discovery. IBM J. Res. Dev. **62**(6), 6:1–6:20 (2018). https://ieeexplore.ieee.org/document/8585034
17. Blunt, N.S., et al.: A perspective on the current state-of-the-art of quantum computing for drug discovery applications (2022). https://arxiv.org/abs/2206.00551
18. Lordi, V., Nichol, J.M.: Advances and opportunities in materials science for scalable quantum computing. MRS Bull. **46**(7), 589–595 (2021). https://doi.org/10.1557/s43577-021-00133-0
19. Cao, Y., et al.: Quantum chemistry in the age of quantum computing. Chem. Rev. **119**(19), 10856–10915 (2019). https://doi.org/10.1021/acs.chemrev.8b00803
20. https://qiskit.org/ecosystem/finance/tutorials/01_portfolio_optimization.html
21. Ostaszewski, M., Trenkwalder, L.M., Masarczyk, W., Scerri, E., Dunjko, V.: Reinforcement learning for optimization of variational quantum circuit architectures (2021). https://arxiv.org/pdf/2103.16089.pdf
22. Tilly, J., et al.: The variational quantum eigensolver: a review of methods and best practices. Phys. Rep. (2022). https://doi.org/10.1016/j.physrep.2022.08.003

23. Fedorov, D.A., Peng, B., Govind, N., Alexeev, Y.: VQE method: a short survey and recent developments. Materials Theory **6**(1), 1–21 (2021). https://doi.org/10.1186/s41313-021-00032-6
24. https://developer.nvidia.com/cuquantum-sdk
25. https://quantumai.google/cirq/experiments/variational_algorithm

Quantum Computing Based Gate Level Optimizer for Digital Electronics

Ravi Theja Kolluru(✉) and Suman Lata Tripathi

School of Electronics and Electrical Engineering, Lovely Professional University, Phagwara, India

ravithejakolluru.tech@gmail.com, tri.suman78@gmail.com

Abstract. An indispensable procedure in the logical synthesis phase of the VLSI-IC design flow is the optimization of the digital circuit. Additionally, the electronic circuits are represented as Boolean expressions using the conventional logic gates as operators. Using unaltered Boolean expressions for IC design frequently results in unfavorable features in an IC, such as an increase in power, chip area, and cost, which lowers the efficiency. This research focuses on optimizing electronic circuit equations, by reducing the number of gates to achieve the same results as the original equations. In the case of traditional approaches, the complexity for an optimal solution increases considerably with an increase in operands and operator count. On the other hand, quantum computing, with the support of the quantum superposition principle, reduces the intricacy of finding effective answers efficiently. This methodology is not only effective in IC design but also beneficial in the execution of Boolean functions in circuits. Simulations are carried out on Vivado software using Verilog for observing IC performance parameters to establish a comparison between the original and optimized equations. The experimental results demonstrate an improvement in the speed of operations and a decrease in power consumption and resource usage. As a result of a drop in gate count, the space occupied by the IC and cost of production will reduce, benefitting the electronics industry. This paper contributes to the application of quantum computing in electronics manufacturing.

Keywords: Algorithm Optimization · Digital Electronics · IBM Qiskit · Quantum Computing · Technology · VLSI

1 Introduction

Digital electronic circuits and devices are the principal part of the digital realm and designing Integrated Circuits (IC) demands deep comprehension of the specifications, in the view of complex requirements. In the design pattern, the logical synthesis is responsible for transforming design requirements into an optimal representation of Boolean expressions at the gate level. Boolean logical optimization techniques are useful in enhancing the efficiency and performance of an IC by effective function execution [1]. The process of achieving optimal Boolean logic functions, specific to a cost parameter, is termed exact synthesis and the intricacy of arriving at it is determined by the

V. E. Balas et al. (Eds.): IEQIP 2023, LNNS 1029, pp. 206–212, 2024.
https://doi.org/10.1007/978-3-031-61929-8_30

combinational complexity [2]. The exact synthesis methods are categorized into graphical, tabular, and algebraic based on execution type. Karnaugh map (K-map) is a suitable example of graphical method optimization to state that the grouping strategies add complexity for designing computerized algorithms. Additionally, computer-based algorithms are comparatively easier to implement in tabular methods such as using truth tables and algebraic techniques through laws of boolean algebra. These algorithmic implementations are often found to be intricate and time-consuming with higher-order logic [3–5]. Later, the Quine-McCluskey approach for optimizing Boolean logic was introduced, but the method complexity grew exponentially with an increase in literals [6]. The application of deep reinforcement learning algorithms to logical optimization has gained popularity recently, but obtaining a reliable dataset is challenging and could result in a solution that is nearly optimized due to the biased data [7]. While broad range of research mentions the usage of electronics for Quantum circuit formations, this paper concentrates on applying Quantum Computing for electronic circuit manufacturing with a novel methodology that involves minimal ancilla qubits by reusing qubits. In addition, only two input gates are used, adhering to electronic gate restrictions. The current research in quantum and nano-technology has enabled the possibilities of exploring logical optimization [8]. Quantum Computing is one of the trending and cutting-edge technologies working on the principles of quantum mechanics and has diverse applications in computation, cryptography, communication, sensing, etc. The computational strength of quantum computing is tremendous and is considered an effective solution for solving optimization problems [9]. The superposition capability of quantum allows it to accommodate higher-order logic, handling both complexity and speed of computation.

This paper summarizes the importance of logical synthesis and the role quantum computing plays in tackling computational complexity for exact synthesis. The results are coupled with simulation for analyzing the IC design parameters. The next Sect. 2 introduces the algorithmic details and simulation procedures used in this study. It also discusses the simulation metrics and the specifications of the software and programming languages employed. Section 3 illustrates the design implementation details of the algorithm and simulation. The results obtained and their corresponding analysis is presented in Sect. 4. The paper concludes by mentioning targets achieved, future perspectives and applications in Sect. 5.

2 Methodology

This research aims to use quantum computing to optimize electronic circuit equations at the gate level. The quantum computer differentiates itself from traditional computers through the fundamentals of superposition and entanglement. Superposition is the ability of qubits, to be in a state where it's both one and zero simultaneously with certain probabilities, mathematically adding up to one [10]. For this reason, unlike the conventional logic optimization algorithms used in VLSI designing, quantum computing provides a computational advantage in optimizing the Boolean equations with higher order by displaying the outcomes simultaneously rather than one at a time [11]. The cost criteria of exact synthesis is aimed at reducing the number of gates used in logic optimization equations and during optimization it is essential to conserve the properties of the input

Boolean expression by ensuring input and output relations are precisely maintained. To fulfill the input-output criteria, the original digital circuit equation is expressed in the form of a truth table, which serves as an input to the quantum computing algorithm. Sequences of logic gates built on a quantum circuit are used to process input variables before being compared to the truth table. The process continues till the optimal answer is obtained and the output is expressed as a Boolean algebra expression. The quantum algorithm operates on Qiskit (v0.43.2) and Python (v3.11.6) programming languages, and uses brute force methodology of passing through all probable scenarios in search of the suitable answer because there isn't any specific procedure for optimization (Figs. 1 and 2).

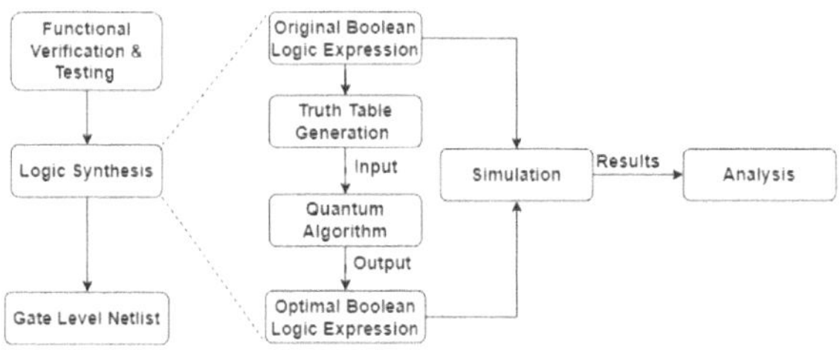

Fig. 1. Approach for gate level optimization in digital electronics using Quantum Computing.

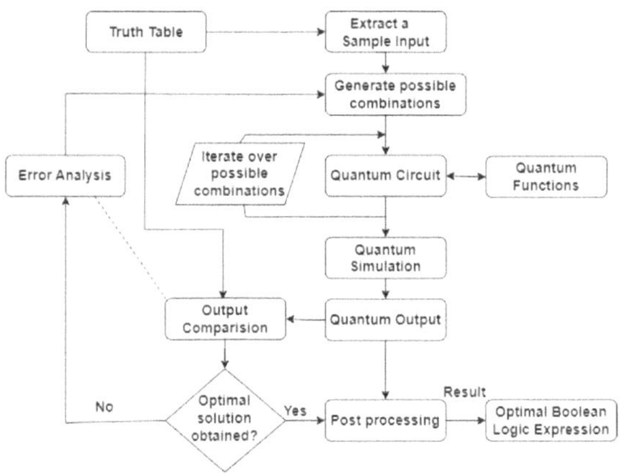

Fig. 2. Block diagram of quantum computing algorithm.

3 Design and Simulation

The design begins with the generation of truth-tables for the original digital circuit equations, as expressed in Boolean functions (f1, f2, f3, f4), as it is well delineated and gives a clear picture of the desired outputs for defined inputs [12]. While the feed-in for the computation is the truth table, the Quantum algorithm iterates over every possible combination of equations, built based on variables and logic gates, using the quantum circuit. Based on the deviation pattern observed from the truth table, the viable outcomes are fed back into the circuit with further logical correlations as part of error analysis. Using the quantum computing, the boolean logic gates (and, or, xor & not gates) are created to represent the proposed boolean expression, as the quantum circuit [13]. In continuation, the quantum circuit is formed using Hadamard, CNOT, Toffoli, SWAP, Fredkin quantum gates, and quantum operations namely measurement, reset, and conditional operations are put to use. The implementation of logic gates by employing quantum gates are labeled as quantum functions. Additionally, a minimum number of ancilla qubits are utilized by making use of swap gate functionality. The optimal boolean logic equations are extracted by the post-processing component from the results of the quantum simulation. Furthermore, by utilizing Vivado's design synthesis function, power analysis is obtained along with resource utilization [14–16]. In continuation, as part of the timing constraints for STA, the input and output delays were configured with 0.5 ns as minimum and 1 ns as maximum timings, with 10 ns as the clock period.

$$f1 = \sum m(2, 3, 4, 5, 6, 7) \tag{1}$$

$$f2 = \sum m(5, 6, 7) \tag{2}$$

$$f3 = \sum m(1, 2, 4, 7) \tag{3}$$

$$f4 = \sum m(3, 5, 6, 7) \tag{4}$$

4 Result Analysis

Comprehensive results from the quantum algorithm reveal that the objective of establishing gate-level optimization for the compact synthesis of digital electronic circuits using quantum computing is achieved. In the process, the reduction in gate count indicates that the cost requirements for logical synthesis were satisfied.

Furthermore, the boolean algebraic expressions initializing the quantum algorithm are weighed up against the results from the algorithm and are displayed in Table 1 along with the gate count for each expression. It's evident from the outcomes that it's possible to attain multiple optimized equations (if any) with the same number of gates, depending on the input equation. As indicated in function-4 (f-4) it is also possible to observe the same expression on both sides with the same tally in gates, usually caused by passing of optimized boolean expression as input. Additionally, if any other expressions match the gate count, they will be displayed as well (Table 1).

The optimal and original equations are transferred to Vivado for analysis where VLSI synthesis and implementation specifications are observed, as shown in Table 2. The analysis of IC design parameters is done on function-1 (f1) from Table-1 for relative comparison. Based on the simulation results obtained, a drop in resource utilization is observed from 13 to 1 for Look Up Table (LUT) and 4 to 3 for IO component count.

Table 1. Original and optimal boolean logic expression comparison

Fn	Original Boolean Logic Expression	Gate Count	Optimal Boolean Logic Expression	Gate Count
f1	(a&~b&~c) + (a&b&~c) + (a&b&c) + (a&~b&c) + (~a&b&~c) + (~a&b&c)	17	a + b	1
f2	(a&b) + (a&c)	3	(a)&(b + c)	2
f3	(~a&~b&c) + (~a&b&~c) + (a&~b&~c) + (a&b&c)	11	a^b^c	2
f4	(a&b) + (b&c) + (a&c); or (a)&(b + c) + (b&c);	5	(a)&(b + c) + (b&c); (b)&(a + c) + (a&c); (c)&(a + b) + (a&b)	4

Meanwhile, it is also important to note that the optimal expression might still require all the defined input and output components causing the IO count to remain constant. The on-chip power, including static and dynamic power, consumed by both the equations are comparatively similar in numbers, the reason being the on-chip power represents the power consumed by the Artix 7 product for simulating the logic equations in AMD Vivado (v2023.1). In the case of STA, while there is a slight decrease in WHS timing, an increase in WNS timing is observed, which are desirable traits. A positive quantity of WNS and WHS conveys that the timing constraints are met and a higher value indicates a greater timing margin implying better design flexibility. Furthermore, as displayed in Fig. 3, the bar chart compares the original expression stated as the sum of products (SOP) and optimized equations in terms of the number of logic gates involved.

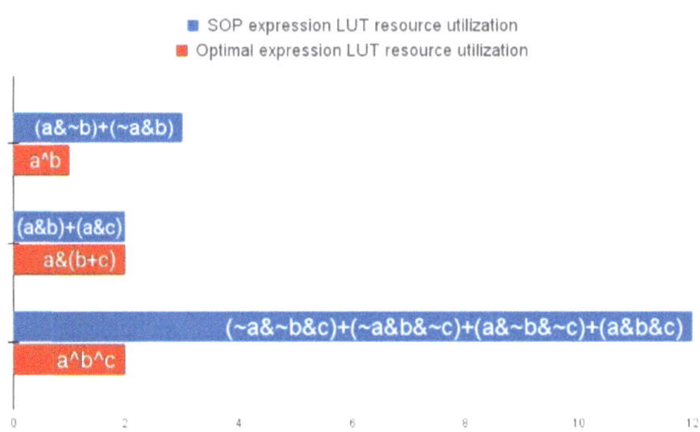

Fig. 3. Comparative display of resources utilization

Table 2. Analysis of IC design parameter comparison for original and optimal expressions.

Parameters	Original Expression	Optimal Expression
On-chip power (mW)	248	246
Dynamic Power (mW)	4	3
Static Power (mW)	244	243
LUT Resource Utilization	20/13	1
IO Resource Utilization	4	3
WNS (ns)	0.332	2.319
WHD (ns)	3.445	3.004

5 Conclusion

The current studies in nanotechnology and quantum have fascinated enthusiasts in exploring their applications for diverse fields. Exact synthesis is an essential process in VLSI IC designing, and the optimization complexity and constraints of traditional logic synthesis algorithms increase with higher-order equations. This research emphasizes the application of quantum computing in the domain of digital electronics in optimizing logical synthesis while dealing with combinational complexity. The Quantum Algorithm used for logic synthesis optimization returns multiple optimized Boolean circuit equations simultaneously. The algorithm is scalable for higher order circuits with ongoing advancements in Quantum by accommodating more qubits onto hardware. The decrease in gate count caused due to the algorithm has led to desirable IC characteristics such as lower on-chip power consumption, resource utilization and static response time. This study has achieved its goal of constructing optimized digital circuits with boosted performance and enhanced efficiency. Moreover, the research contributes to the domain of electronics by enabling an increase in computational speed, reduction in size and manufacturing cost adding value to the industry and end users.

References

1. Palnitkar, S.: Verilog HDL: A Guide to Digital Design and Synthesis, 2nd edn. SunSoft Press, India (2003)
2. Soeken, M., et al.: Practical exact synthesis. In: 2018 Design, Automation & Test in Europe Conference & Exhibition (DATE), pp. 309–314. IEEE, Germany (2018)
3. MacFarlane, A., Russell-Rose, T., Shokraneh, F.: Search strategy formulation for systematic reviews: issues, challenges and opportunities. Intell. Syst. Appl. (15), 200091 (2022)
4. Pandit, V., Schuller, B.: A novel graphical technique for combinational logic representation and optimization. Complexity (2017)
5. Miller, D., Nadathur, G.: Programming with Higher-Order Logic. Cambridge University Press, USA (2012)
6. Jain, T.K., Kushwaha, D.S., Misra, A.K.: Optimization of the Quine-Mccluskey method for the minimization of the Boolean expressions. In: Fourth International Conference on Autonomic and Autonomous Systems (ICAS 2008), pp. 165–168. IEEE, France (2008)

7. Haaswijk, W., et al.: Deep learning for logic optimization algorithms. In: 2018 IEEE International Symposium on Circuits and Systems (ISCAS), pp. 1–4. IEEE, Italy (2018)
8. Amarú, L., Vuillod, P., Luo, J., Olson, J.: Logic optimization and synthesis: Trends and directions in industry. In: Design, Automation & Test in Europe Conference & Exhibition (DATE), pp. 1303–1305. IEEE, Switzerland (2017)
9. Primqulov, O.: The pursuit of quantum supremacy: challenges and Implications. Innov. Dev. Educ. Act. **2**(7), 200–205 (2023)
10. Rieffel, E., Polak, W.: An introduction to quantum computing for non-physicists. ACM Comput. Surv. (CSUR) **32**(3), 300–335 (2000)
11. Soeken, M., Wille, R., Keszocze, O., Miller, D.M., Drechsler, R.: Embedding of large Boolean functions for reversible logic. ACM J. Emerg. Technol. Comput. Syst. (JETC) **12**(4), 1–26 (2016)
12. Drechsler, R., Wille, R.: From truth tables to programming languages: progress in the design of reversible circuits. In: 41st IEEE International Symposium on Multiple-Valued Logic, pp. 78–85. IEEE, Finland (2011)
13. Liu, H., Li, F., Fan, Y.: Optimizing the quantum circuit for solving boolean equations based on Grover search algorithm. Electronics **11**(15), 2467 (2022)
14. Tripathi, S.L., Saxena, S., Sinha, S.K., Patel, G.S.: Digital VLSI Design and Simulation with Verilog. Wiley (2021)
15. Tripathi, S. L., Prasanna, D. L., Mahmud, M.: A Low-Power Audio Processing Using Machine Learning Module on FPGA and Applications. Explainable Machine Learning Models and Architectures. Wiley (2023)
16. Badiganti, P. K., Peddirsi, S., Rupesh, A. T. J., & Tripathi, S. L.: Design and implementation of smart healthcare monitoring system using FPGA. In: Rawat, S., Kumar, A., Kumar, P., Anguera, J. (eds.) Proceedings of First International Conference on Computational Electronics for Wireless Communications. LNNS, vol. 329, pp. 205–213. Springer, Singapore (2022). https://doi.org/10.1007/978-981-16-6246-1_18

Graph Optimization Studies with QAOA Using Pennylane

Alhajie Mohamed Sesay[1], S. Guru Devanur[2], Thinagaran Perumal[3], and Kolla Bhanu Prakash[1](✉)

[1] Department of Computer Science and Engineering, Koneru Lakshmaiah Education Foundation, Guntur, Andhra Pradesh, India
drkbp@kluniversity.in
[2] Mysore University, Mysore, India
[3] UPM Malaysia, Serdang, Malaysia
thinagaran@upm.edu.my

Abstract. Graph optimization problems are used in many different fields, including machine learning, network architecture, and logistics. They often require determining the best way to arrange variables in a way that minimizes or maximizes a certain objective function. For traditional algorithms, these problems are known to be computationally demanding, particularly for large-scale graphs. A promising approach to solving these challenging optimization issues more effectively is provided by quantum computing. The Quantum Approximate Optimization Algorithm (QAOA) is a quantum algorithm specifically designed to handle combinatorial optimization challenges, including graph-related tasks. QAOA iteratively improves the quantum state by varying the circuit's parameters, which approximates the ideal solution. Quantum algorithms can be tailored to address specific graph-related tasks, such as maximum-cut, graph partitioning, and clique finding. QAOA promises to revolutionize the field of graph optimization by providing exponential speedup compared to classical algorithms. However, noise and error mitigation challenges in current quantum hardware remain critical obstacles to fully realizing the potential of QAOA for large-scale graph problems. Nevertheless, ongoing research and technological advancements in quantum computing are expected to address these limitations and pave the way for exciting new possibilities in graph optimization with QAOA. Overall, this abstract serves as an introduction to the emerging field of graph optimization with QAOA, providing an overview of the fundamental concepts, potential applications, and the challenges ahead.

Keywords: Graph Optimization · QAOA · Quantum Computing · Quantum Circuit · Pennylane

1 Introduction

Graph optimization is a fundamental problem with numerous real-world applications across various domains. It involves finding optimal solutions for tasks such as finding the maximum cut in a network, identifying the most efficient routing paths, optimizing

V. E. Balas et al. (Eds.): IEQIP 2023, LNNS 1029, pp. 213–221, 2024.
https://doi.org/10.1007/978-3-031-61929-8_31

resource allocation, or clustering nodes with similar characteristics in a social network. These problems are computationally difficult to solve because their complexity increases exponentially with graph size. Conventional methods for graph optimization frequently depend on classical algorithms, which can have drawbacks when dealing with complicated, large-scale issues. Also, to examine are the essential elements and procedures of implementing QSOA, such as setting up quantum registers, building the QSOA circuit, and doing classical circuit optimization. Choosing and creating objective functions that are appropriate for graph optimization is a challenging problem.

The prominent graph optimization applications of QAOA include network clustering, maximum cut, graph coloring, and travelling salesperson issues. The advantages and difficulties of applying QAOA to various applications include scalability, available quantum resources, and problem size. Advancements in circuit design, variational methods, and noise reduction strategies targeted at improving QAOA performance and dependability for more complicated and large-scale graph optimization issues are bothering in present day research.

QAOA is based on mapping an optimization problem to a quantum Hamiltonian, a mathematical operator used in quantum mechanics to describe the energy of a quantum system. By designing a Hamiltonian that encodes the optimization problem's objective function, QAOA aims to find the lowest energy state of the Hamiltonian, which corresponds to the optimal solution of the optimization problem. QAOA uses variational quantum circuits, sequence of quantum gates parameterized by certain angles. These angles are treated as parameters to be optimized. By varying these angles, the quantum circuit can investigate many states and progressively gets closer to the ideal answer. Classical optimization procedures are utilized to optimize these angles.

Single-qubit rotations and entangling gates are the two primary categories of quantum gates that make up each layer in QAOA. The continuous transformation created by these gates modifies the quantum state of the system. The main objective of QAOA is to increase the Hamiltonian's expectation value. Determining quantum circuits, optimizing parameters, and interacting with quantum devices or simulations are all made easy using PennyLane.

2 Literature Review

The aim of the Quantum approximation Optimization Algorithm (QAOA) is to solve approximation combinatorial optimization problems as fast as possible. Combinatorial optimization problems aim to find the optimal configuration or combination of components from a discrete set to optimize a given objective function [4, 5]. These problems are widespread in several fields, including cryptography, graph theory, scheduling, physics, and logistics [7, 8]. Quantum Approximate Optimization Algorithm for Maximum Cut by Hadfield, Jiang, and Sanders; explains how QAOA performed on the Maximum Cut problem, a key graph optimization challenge.

Gao, Z., Li, Y., & Duan, introduced Quantum Graph Clustering. The authors suggest using a quantum method to determine optimal clustering of graph nodes and analyzing its performance on synthetic and real-world graphs [9, 10]. The goal of combinatorial optimization issues is finding the ideal configuration or combination of discrete parts to optimize a given objective function. In contrast to classical algorithms, the quantum algorithm known as QAOA uses quantum superposition and entanglement to investigate alternative solutions to combinatorial optimization issues [1]. The authors discuss the potential advantages of quantum algorithms over classical methods and highlight the challenges in scaling QAOA to solve large-scale graph problems [2, 3].

The influence of the number of layers in the QAOA circuit and its impact on the quality of solutions is very important. Quantum Approximate Optimization Algorithm and its use to solve graph optimization issues are thoroughly covered in these references. Researchers interested in this field can refer to these works to understand the theoretical foundations, implementation challenges, and potential applications of QAOA in solving real-world graph optimization tasks [6].

3 Methodology and Building Blocks of QAOA

Graph optimization is one application where near-term quantum hardware is anticipated to excel. Because of their frequent complexity and close ties to real-world applications (such as logistics and social networks), graph-based problems are intriguing to investigate. Graphs are made up of several connected nodes. Cost of Hamiltonian and accompanying mixer Hamiltonian are first calculated to solve using QAOA in PennyLane and Amazon AWS Quantum Braket. This will give us binary samples that allow us to select which graph nodes to use as part of our clique, e.g., by simply selecting the most common piece or selecting a sample with the lowest corresponding energy (Fig. 1).

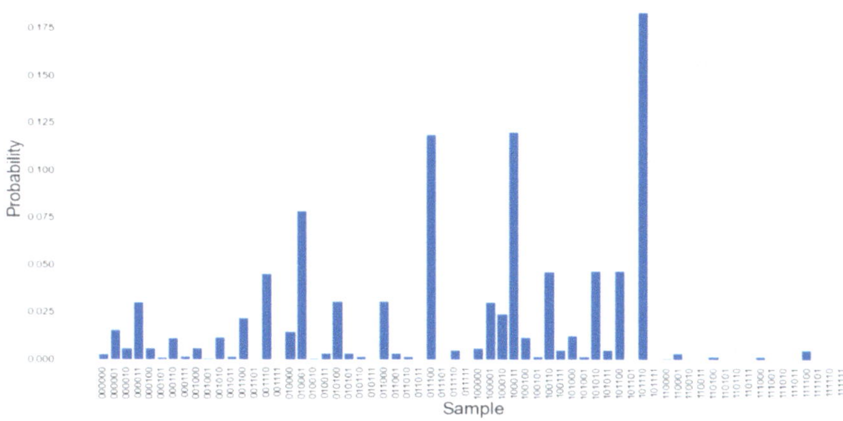

Fig. 1. Plot showing the probability distribution over all possible samples.

From the plot, the sample 101110 has the most significant probability. Since each qubit corresponds to a node, this sample selects the nodes [0, 2, 3, 4] to form a subgraph. QAOA works by finding the optimal solution through quantum evolution. We need to make sure that the problem can be formulated in a way that takes advantage of QAOA's optimization capabilities. For large graphs, you might need to partition the problem into smaller sub-problems that can be solved more easily. This can help reduce the complexity of the computation and make it more amenable to quantum processing. Hardware constraints, depth & parameters, noise-error mitigation, and different algorithm variations need to be investigated. Distributed and parallel approaches with high scalability is the target to achieve using quantum computing (Fig. 2).

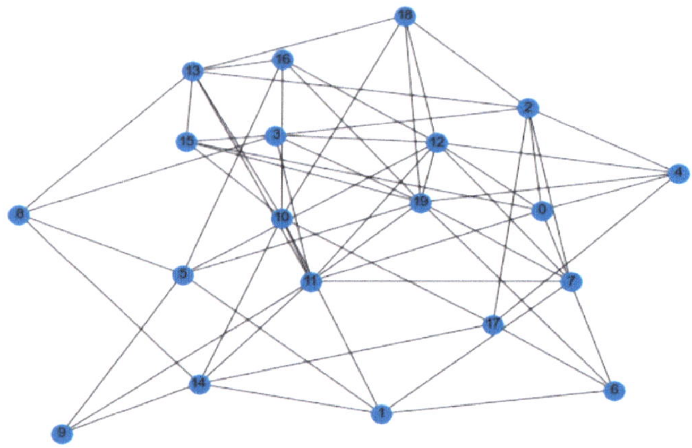

Fig. 2. Optimization problem on a twenty-node graph

A twenty-node graph puts us in a regime where the local simulator will be slow to execute. The maximum cut problem must be solved to split the graph's nodes into two groups that share the most significant number of edges. According to NP-hard, we anticipate that it will become more challenging as the number of graph nodes rises. The QAOA algorithm is a member of the group of hybrid quantum algorithms, or Algorithms known as NISQs, or noisy intermediate-scale quantum, are regarded as the mainstays of the NISQ era now. The emerging method for benchmarking quantum devices in the NISQ era is a leading prospect for demonstrating a real-world quantum speed-up on a near-term NISQ device, known as QAOA [1, 4]. We compare our results to those obtained from traditional QUBO solvers to validate our methodology.

The training of deep neural networks uses an objective function that you wish to reduce, which is commonly represented as the error on your training set. First begin with an initial approximation for the weights in the network to reduce that error. The coprocessor, in that case a GPU, takes these weights which define the exact operation to execute, and the output of the neural network is computed. This output is then used to calculate the objective function, which is then utilized by the CPU to make an choice to update the weights and the cycle continues. Variational quantum algorithms, a specific form of hybrid algorithms, work in the very same way, using parametrized quantum course rather than parametrized neural networks and replacing the GPU with a QPU. Here, you start with an initial guess for the parameters that define your circuit, have the QPU execute that circuit, perform measurements to calculate an objective function, pass this value back to the CPU and have this classical CPU update the parameters based on that information. Quantum computers are single threaded, often located in specialized lab settings, and have unique latency needs. The undifferentiated heavy lifting is precisely what Amazon Braket does for us so that we may concentrate on our scientific issues.

Combinatorial optimization problems are ubiquitous across many areas of science and application areas. Applications can be found in physics, mechanical engineering, logistics, scheduling, portfolio optimization and planning. Combinatorial optimization problems, in a nutshell, are issues containing a lot of yes/no options, with each set of decisions producing a value for an associated objective function, such as a cost or profit value. A wide range of NP-hard combinatorial optimization issues are unified by the QUBO (Quadratic Unconstrained Binary Optimization) model. Famous instances include the Maximum-Cut Problem, Task Allocation Problems, Capital Budgeting Problems, and Quadratic Assignment Problems. Among the class of QUBO problems, Maximum Cut (MaxCut) is a challenge in combinatorial optimization. Specifically, we can encode the Maximum Cut problem in physics as a minimization problem of an Ising Hamiltonian.

Since we are primarily interested in solving the classical optimization problem, within this routine it is sufficient to keep track of the best classical bitstring. This means that the wavefunction prepared by the quantum circuit must have some overlap with the optimal solution that we can read out as bitstring in the measurement shots. In principle (i.e., without any training), we could just sample from an utterly uniform state that is prepared in a superposition of all computational basis states, as prepared by applying Hadamard gates to all qubits. For large systems, however, this approach is not scalable as we would need to take an exponentially improving number of measurements. That is why we train our circuits, update the parameters, with the goal to increase our success chances to find the optimal bitstring. Consider Using the cost Hamiltonian excited low energy eigenstates as an example, a variational state other than the ground state $|z*\rangle$. This state will have a reasonably low expectation value $\langle \hat{H}_C \rangle$ As this low energy state does not coincide with the ground state, the success chance is zero. The same is true for a variational form that is a linear combination of the ground state with extremely high energy eigenstates (γ, β), while reporting a high cost value to the classical optimizer. The best option is to use the Gibbs objective function (Fig. 3).

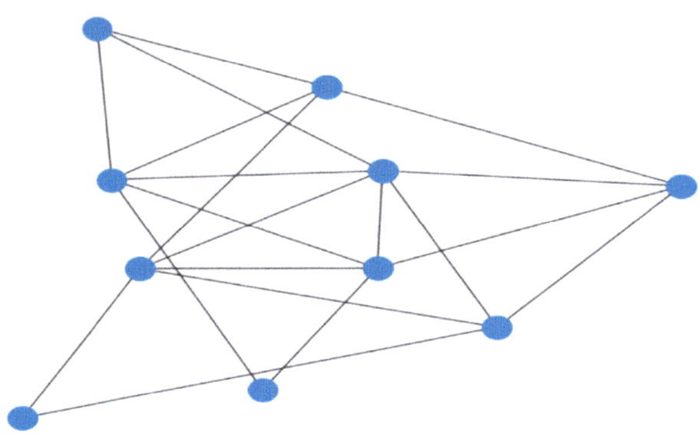

Fig. 3. Erdos Renyi graph

Given a graph G = (V,E), consisting of a particular set of vertices (also known as nodes) V and edges E, our goal is to color each node red or blue, then score a point for each node next to a node of different color. We strive to find the optimal coloring that achieves the most significant number of points. To this end, we will address the dual challenge of determining the equivalent Ising Hamiltonian's minimum energy. To get started, we first use the open-source network library to visualize the problem graph (Fig. 4).

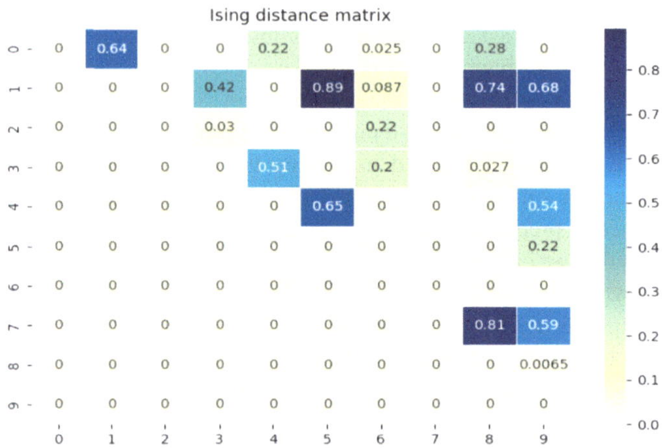

Fig. 4. Plot showing Ising distance matrix

4 QAOA Implementation, Results and Discussion

The main building blocks for the core modules of our QAOA algorithm are (i) a function called circuit that defines the parametrized ansatz, (ii) a function called objective_function that takes a list of variational parameters as input, and returns the cost associated with those parameters and finally (iii) a function train to run the entire QAOA algorithm for given ansatz. This way we can solve the problem in a clean and modular approach. We are now all set to run some QAOA simulation experiments. First, try with the number of qubits N. Secondly, you may also test with the classical optimizer. Since we use an off-the-shelf, black-box scipy minimizer, we can simply swap between different optimizers by setting the OPT_METHOD parameter. Some popular options readily available within this library include Nelder-Mead, BFGS and COBYLA. As a precautionary warning, note that the classical optimization step may get stuck in a local optimum, rather than finding the global minimum for our parametrized QAOA ansatz wavefunction. To address this issue, we may run several optimization loops, starting from different random parameter seeds. While this brute-force approach does not provide any guarantee to find the global optimum, from a pragmatic point of view at least it does increase the odds of finding an acceptable solution, at the expense of potentially having to run many more circuits on the simulator or QPU, respectively. Finally, the optimization loop may require the execution of many individual quantum tasks (i.e., single circuit executions for fixed parameters). For example, when choosing the classical Powell optimizer for the graph considered here, we find ~270 cycles in the for loop. For the local simulator device selected here by default this is not an issue, still, if you run this algorithm on any QPU you may want to adjust the maxfev parameter to control the maximum allowed number function evaluations.

Specifically, we display the results for the variational parameters β and γ for every layer in our QAOA ansatz. Moreover, we show the solution to our graph coloring problem with every node colored red or blue. Finally, we compare these results to results found classically using the open-source pyqubo package. For the original small toy problem we have set up, there are several degenerate classical solutions with the same optimal quality. The classical and the QAOA approach may find solutions with different coloring configurations but the same quality (that is energy). Secondly, with QAOA we are not guaranteed to find optimal solutions. Specifically, the deeper the circuit, the more complex the classical optimization problem, and we may get stuck in a local rather than global optimum. One brute-force approach is then to re-run QAOA with different random initial seeds for the parameters (β, γ) (Fig. 5).

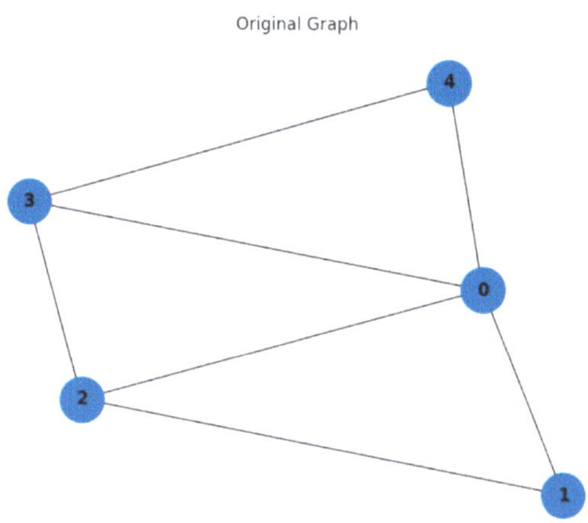

Fig. 5. Quantum Approximate Optimization Algorithm (QAOA) optimization on that graph

QAOA may arrive at a different solution than our classical benchmark code. Firstly, the classical optimization routine may get stuck in a local rather than global optimum. To avoid this, more sophisticated optimization strategies may be employed in future. Secondly, even if QAOA arrives at the same classical energy as our classical approach, the coloring may differ since the solution space may be degenerate for the specific example shown here (this means, two different classical bitstrings have the same energy). At minimum, you may find an inverted coloring (by swapping red and blue colors), because of the underlying Z2 symmetry.

5 Conclusion

To implement quantum universal gates on larger qubit equivalent system using graph theory for optimization studies, we can change the probability amplitude of the targeted closed set. Taking help of conditional position operators, hybrid quantum ensemble learning algorithms can be executed. Executing quantum algorithms like Shor's algorithm, Grover's search, quantum Fourier transform, QAOA and quantum phase estimation for three-qubit equivalent system is another challenge. A comparison of circuit complexity and circuit depth shows that the proposed quantum walk scheme reduces the complexity when compared to circuit model in many cases. However, with a careful designing of position dependent evolution operators one can execute the implementation of various quantum computational tasks (Fig. 6).

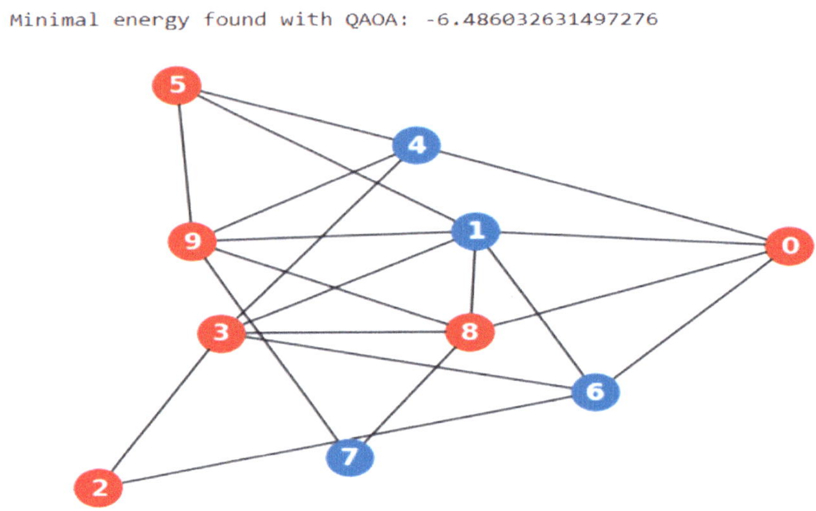

Fig. 6. Graph of Minimal energy found with QAOA

References

1. Farhi, E., Goldstone, J., Gutmann, S.: A quantum approximate optimization algorithm. arXiv: 1411.4028 (2014)
2. Cao, Y., et al.: Quantum chemistry in the age of quantum computing. Chem. Rev. **119**, 10856 (2019)
3. Smith, A., Kim, M., Pollmann, F., Knolle, J.: Simulating quantum many-body dynamics on a current digital quantum computer. NPJ Quantum Inf. **5**, 1 (2019)
4. Zhou, L., Wang, S.-T., Choi, S., Pichler, H., Lukin, M.D.: Quantum approximate optimization algorithm: performance, mechanism,and implementation on near-term devices. arXiv:1812. 01041 (2018)
5. Glover, F., Kochenberger, G.: A tutorial on formulating and using QUBO models. arXiv: 1811.11538 (2018)

6. Vikstal, P., Groenkvist, M., Svensson, M., Andersson, M., Johansson, G., Ferrini, G.: Applying the quantum approximate optimization algorithm to the tail assignment problem. arXiv:1912. 10499 (2019)
7. Li, L., Fan, M., Coram, M., Riley, P., Leichenauer, S.: Quantum optimization with a novel Gibbs objective function and ansatz architecture search. arXiv:1909.07621 (2019)
8. Farhi, E., Goldstone, J., Gutmann, S.: A Quantum approximate optimization algorithm. arXiv preprint arXiv:1411.4028 (2014)
9. Hadfield, S., Jiang, Z., Sanders, B.C.: Quantum approximate optimization algorithm for maximum cut: a fermionic view. arXiv preprint arXiv:1906.05821 (2019)
10. Wang, Q., Cao, Y., Chen, P.: Quantum approximate optimization algorithm for graph partitioning. Quantum Inf. Process. **19**(11), 421 (2020)

Author Index

V. E. Balas et al. (Eds.): IEQIP 2023, LNNS 1029, pp. 223–224, 2024.
https://doi.org/10.1007/978-3-031-61929-8

SPRINGER NATURE

GPSR Compliance

The European Union's (EU) General Product Safety Regulation (GPSR) is a set of rules that requires consumer products to be safe and our obligations to ensure this.

If you have any concerns about our products, you can contact us on ProductSafety@springernature.com

In case Publisher is established outside the EU, the EU authorized representative is:

Springer Nature Customer Service Center GmbH
Europaplatz 3
69115 Heidelberg, Germany

The manufacturer's authorised representative in the EU is Springer
Nature Customer Service Centre GmbH, Europaplatz 3, 69115 Heidelberg,
Germany. If you have any concerns regarding our products, please
contact ProductSafety@springernature.com

Printed and bound by CPI Group (UK) Ltd, Croydon, CR0 4YY
23/09/2026
02221527-0003